Our Existence

Pt 1

The Nature and Origin of Physical Reality

A straightforward and impartial guide to
the science and philosophy of
our physical environment

Christophe Finipolscie

Published by   Spontaneous Publications Ltd.,
                   Charter House, Marlborough Park, Southdown Road,
                   Harpenden, Herts., UK AL5 1NL
                   info@spontaneouspublications.com

Wherever possible the publishers have sought copyright approval for all images / photographs used, but are happy to adapt the known attributions if other valid claims can be established.

This book is sold subject to the condition that it shall not by way of trade or otherwise, be lent, resold, hired out, or otherwise circulated without the publisher's prior written consent, in any form, binding, or cover other than that in which it is published and without similar conditions including this condition, being imposed on the subsequent purchaser.

First Edition November 2016
First published February 2017 – Imprint – Spontaneous Education

ISBN   978-0-9956491-0-1 standard Hardback full colour edition.

ISBN   978-0-9956491-1-8 standard Paperback full colour edition.

ISBN   978-0-9956491-4-9 primary electronic edition

A CIP catalogue record for this book is available from the British Library

This book is produced and published under the laws of the United Kingdom of Great Britain and Northern Ireland.

CONTENTS

Foreword & Acknowledgements

This book, and the others in this series, have been in development for over 20years, and in that time many people have read occasional chapters and shown their enthusiasm and encouragement, which has kept me going.

Sadly, there are too many to sensibly list here, but I can assure them that their comments were taken 'on-board' and that the time they donated in reading a partly formed work was greatly appreciated.

However the efforts made by some people in reading and sometimes re-reading the whole book in its various stages, need special recognition. Inevitably my wife deserves a first mention as she had to put up with many lonely evenings while I wrote these works, and without her tolerance these books would still be an idea rather than a reality. In addition, my special thanks have to go to the following people for their patient reviews and commentaries across some or all of these books :-

> Dr. A.Punter, Mr. J.Francis, Mr. N.Dyer, Mrs. C.Wilmot

I can't thank you enough for all your efforts.

As this book is intended to reflect all perspectives I would like to make it clear that neither I nor any of the reviewers endorse any particular viewpoint, but we have tried to fairly reflect the different opinions.

Although they have asked to remain anonymous I would also like to extend particular thanks to a small number of other specialists who have validated different sections and in their different ways have also spurred me on, as the three quotes below will hopefully illustrate:

Senior Scientist 1
"These are great ideas and they need to be told. Some of the things you point out have never occurred to me before but they do have a strong ring of truth about them...   It's good to be given a fresh perspective on occasions, but you'll appreciate that people in my position can't easily venture into this territory without it becoming career limiting!"

Senior Scientist 2
"If we set aside some nit-picking I can't deny what you say about Quantum Mechanics but I don't really like you saying it. It isn't how I like to think about the subject... having said that, it is interesting to see the different ideas laid out side by side ... and I do see why you've put them into a philosophical framework, but I remain a committed Materialist at heart."

Theologian
"People of faith cannot deny the advances that science has made, so they have to find ways in which to reconcile them with their various beliefs. This is the first book I have read which has truly given me a basis for doing this – and you've done it in ways that anyone can understand."

Christophe Finipolscie
November 2016

1 Setting the Scene

From the moment that we wake to the time we fall asleep our physical environment dominates our experience of life, as it has done for all the generations before us. It's a dynamic existence that is largely predictable, and this allows us to use the materials within it for our own purposes, to build homes, vehicles, gadgets, and things of beauty.

In day-to-day terms that familiarity and control leads most of us to think that we know what this 'stuff' is. *Yet we don't.* At its core we <u>don't</u> know what physical matter is made of. There is still a mystery to be solved… although there are a few logical suggestions that match the evidence - some of which may surprise you.

The huge advances in our knowledge over the past two centuries have certainly allowed us to get a much deeper insight into our circumstances than our distant ancestors. However, scientific discoveries over the past four decades have once again been **fundamentally** changing perceptions about what things are and how they arose – even to the extent that mainstream theories such as 'Big Bang - Big Crunch' now face serious challenges, as we will see later.

It's a good time to re-appraise what we know.

Ever since modern man first walked the Earth we have wondered what we are and where we came from. This book is part of a short series that will help you form an understanding of our existence: setting the latest scientific findings alongside the full range of philosophies which help us interpret facts and speculate about the unknown.

We see ourselves as physical beings, so physical reality will inevitably form part of our story, and yet we see a sharp distinction between living things and the array of sterile chemicals which make up our environment. This volume focuses on the physical aspects of our reality, while the processes of Life and the more disputed topics of the Mind and Soul form other parts of this series.

> Inevitably, these books share some common themes, because they are closely linked. To this extent the first couple of chapters of each book will introduce basic concepts that underpin all lines of enquiry even if they are presented with a slightly different slant. However later chapters will be quite distinct.

While the expertise of science is to determine facts, those findings rarely lead to *understanding* by themselves. Facts have to be interpreted before possible answers can be revealed. We also know that there are still many gaps in our knowledge, which inevitably leads to speculation, especially from scientists who discover new things. This is where people enter the realm and the expertise of philosophy.

Philosophy isn't religion. It shows all of the different viewpoints from which facts can be interpreted, based on 'abstract' logic rather than doctrine. For this reason, both scientific and religious perspectives can find their place within it. Its reasoning places a framework around the unknown and to this extent, science and philosophy should work together to improve our understanding.

When I set out on my own journey to learn about the reality of existence I was surprised at how difficult it is to establish the facts that are relevant to this debate. We not only have to battle against the narrow perceptions of headlines and the barriers created by jargon, but also against a culture within science that discourages philosophical interpretation. It is therefore very hard to know when a discovery may be significant to this discussion because it won't be labelled in that way.

Take the following example of a quote from NASA, which is here to illustrate a point, not to worry anyone, (so I also provide a translation) :-

> *The 9 year results of the WMAP survey[1] concluded that the universe is flat with only a 0.4% margin of error and that Euclidean geometry probably applies.*

Layman's translation : No theory of origin which is purely based on a conventional explosion could produce something of infinite size, including the Big Bang.

If you want a purely physical explanation for the origin of existence then limiting the size of the Universe becomes quite important, and suggesting that the 'fabric of space' is curved has been one way to do this.

If you imagine space, (the cosmos), being laid out in a grid with arrows pointing in the directions of height, width, and depth you may either imagine those lines remaining straight into infinity, or you might imagine them having a subtle curve.

Any curve might ultimately bend round on itself to form a circle or ball, and in the context of the cosmos a curvature of space might well mean that the Universe had a limited size.

Well this finding indicates that the 'fabric of space' probably isn't curved, and may therefore be infinite.

I have written this book to help you avoid years of research into obscure areas of mainstream science and philosophy. My aim is to present the main facts in clear language so that everyone can understand the evidence and the issues, which you can then investigate further if you wish.

I am also interested in the full *range* of opinion, so I am not here to promote a cause.

I want to open-up the debate by presenting the different ideas side by side, warts and all, so that we can fully judge the merits of each concept and how they all relate to each other. Although I can only present you with a snapshot of what is known, it will at least enable you to observe and assess the changes in thinking going forwards.

Science is strongest in the subjects of Physical Reality and Physical Life because these are the most tangible and accessible aspects of existence, so the associated academic disciplines have generated more facts to support our perceptions.

The study of the Mind has had less funding and is far less accessible, so it has fewer facts. At the far end of the scale, and despite religious fervour across the centuries, the study of the Soul has produced the least number of facts to justify its existence – but there <u>are</u> some, which have to be explained even if you disagree with the concept.

Facts by their nature are robust, (although they may be perceived and explained in different ways). Other statements will ultimately be shown **not** to be facts due to fraud or speculation, but genuine raw facts don't change – ever. Let me explain further.

If you put something into a process and get a particular result, that is a fact which will not change, and which has to be explained even if we later find that there were errors or other things involved which we were originally unaware of.

We also have to be careful in the way that we phrase things. It would be wrong to claim that a fact has changed because (say) "Pluto is no longer a planet'. Nothing has happened to Pluto. It is the way that we label it that has changed, and the earlier fact still stands; it <u>was</u> labelled that way on the earlier categorization.

So in any particular declaration we have to separate the fact from the interpretation.

An interpretation of experimental results can lead us to a theory about what is going on, and we can be so familiar with an interpretation that people can often mistake it for a fact, when it isn't. In this case, only the experimental results would be a fact.

If there's any reasonable doubt about something it can't be a fact.

Because of this they should all be explained, even if they are inconvenient to a particular theory. I feel it's dishonest to ignore facts that don't fit your theory, and to present only part of the story simply because it suits your preferences. That is true of atheism and materialism as much as it is of religion and dualism, (and if you are not familiar with all of these terms – again, don't worry, all will be explained later).

I say all this because, in relation to some key issues, the real science you will read about in this book is beginning to support old ideas about the nature of existence which reflect more abstract philosophical thinking... and if new ***principles*** are being suggested we have to consider the wider implications as well. Bizarre but true.

Science has a long history of poaching religious or unusual ideas, in support of new evidence. For instance:-

- the notion that our physical reality is made up of atoms first emerged 2,500 years ago in the religious philosophy of Atomism[2].
- The scientific 'Multiverse Theory' has a lot of similarity with the 'other planes of existence' or parallel worlds referred-to in Hinduism.

Just because something is labelled as religious doesn't mean that it's invalid, as long as there are reasoned grounds which match the evidence. Equally, science is not atheism. Its findings point 'both ways', and after centuries of disproving incorrect religious pronouncements, the 'swings of fate' have started to work against some atheistic ideas.

I think we all realise that people can be very passionate about these subjects, and I must therefore ask you not to fall for the smear campaigns of those who wish to discredit other beliefs rather than disprove them. A good idea is a good idea, regardless of where it came from. In ancient times when comparatively little was known about the nature of the physical world, the blind were effectively leading the blind and some very inaccurate and discredited beliefs served to tarnish the reputations of both science and philosophy. For instance, the Earth is not flat, and the sun isn't pulled around it by the chariots of the Gods.

In modern times however, much more is known and proven about our physical environment, so the speculation is limited to quite discreet areas, and the justifications for our theories have to be much more robust.

The fact that many philosophies remain viable today means that they still offer reasonable explanations <u>within the knowledge that we currently have</u>... and this is true of ideas which may sound very strange at first.

I say again that I intend to present all of the perspectives with an even hand to allow you to choose what you feel is correct. But be prepared to be surprised!

While I will try hard to give each perspective a fair hearing, we need to bear in mind that some philosophies have existed for a lot longer than others, and these will tend to have more arguments in their armoury. So if I spend longer discussing some subjects rather than others it will not reflect bias. Equally we can say that if, after many centuries, a philosophy **hasn't** been able to explain a particular issue very well, this may point to a real problem with that perspective.

It is also important to consider the limits of certain arguments and how far they might reasonably be taken. For instance, it's very easy for people to assume that a description is an explanation and that formulae are proof rather than theory.

We should also remember that our knowledge is evolving and that science itself has been shocked on several occasions when it lifted the lid on new aspects of research, to find that it had been wrong and that there were much deeper issues to uncover.

The new facts which were uncovered didn't invalidate the old ones, but they shifted opinion in the way that the earlier facts were interpreted. The emergence of Quantum Mechanics is one of the more recent and most profound examples of this, and it is from here that new and robust evidence has led to very different ideas about the structure of our Universe, which are now taken very seriously.

Don't worry if these words sound daunting, the underlying ideas are straightforward when presented simply – and that is my aim. The basic point that I am making is that there's a *deepening* mystery to be solved about what existence is and how it arose.

This hopefully sets some of the context for the topic, but before we go any further I must explain a convention that I use throughout the book:-

- I will refer to *mechanisms* with a capital letter, (eg. Thought, Time), and
- their *effects* in lower case (eg. our thoughts, or the passage of time).

So let's now focus on the subject at hand: physical reality.

1.1.1 Yardsticks, Mysteries & the Questions we are Trying to Answer

Having mentioned that there's a mystery to be solved about the nature of existence, I feel it is time to illustrate the scale of that mystery.

As I mentioned before, even to this day science still doesn't know what lies at the core of existence. While it **has** discovered that the atoms of different materials are all made from the same building blocks, (known as sub-atomic particles), and these are assembled in different combinations and in different layers… we still don't know :-

- what those components are ultimately made of, (ie. how many **types** of underlying 'stuff' may underpin existence), or
- how they may interact, or even
- whether we have reached the bottom of the many layers of existence that are to be found within atoms.

All we can say is that the activities of the tiniest particles so far discovered, (the deeper levels of existence), break most of the rules which seem to apply to our day to day experiences of physical matter, (at the higher levels). Put another way, different layers of existence seem to have different Laws of Physics.

> For instance, some sub-atomic particles can be shown to hit one side of an 'object' and then disappear, only to re-emerge as if by magic on the other side. It's as if existence is being passed-through things of substance.

> Other observations suggest that a single particle may be in two places at once, or may spin in opposite directions **at the same time** (!) [3.32]… however because we can't directly see such particles, we can't be absolutely sure what factors our devices are actually recording.

The results <u>are</u> facts, but what do they mean? Is it just that our equipment isn't sophisticated enough to let us know the underlying reality?

To this extent, our scientific exploration of physical reality is working inwards towards the core. Yet our dilemma is equally great when we look outwards to try to establish the 'big picture' and the totality of physical existence, because we cannot see the edge of the universe, even with our most powerful telescopes. As mentioned earlier, each new generation of device reveals more and more, and the galaxies which are on the limits of visibility today are a very long way indeed, from what we can tell. The most prominent estimates of the size of the Universe exceed 93 billion light years.

Why is that important? Well, if you compare this to the age of the Universe, (13.77 billion years), even the bits that we *can* see extend many times further than could be achieved by anything travelling at the speed of light, (the scientific **maximum** speed that *anything* could reach). I discuss the size of the universe in Chapter 4.

Put another way, even if the Universe is not infinite, the only way that a huge explosion from a tiny point in space could explain the origin of our Universe is for the Laws of Physics to be broken. We therefore have to ask if this theory is either incomplete in itself; part of a bigger story; or even entirely wrong? Fudging the issue by declaring that the expansion of the Universe is an event outside the Laws of Physics doesn't take away from the core point, and new scientific theories of origin <u>are</u> emerging.

If we can't see any boundaries or limits, either looking outwards or inwards, then we need other fundamental yardsticks by which to gauge our perceptions of reality. Unfortunately, in terms of a 'stake in the ground', there are very few of these.

Perhaps just two. One is a fact, the other is a core belief.

In practical terms, the primary yardstick on which we assess our day-to-day activities is the rigid consistency of how nature seems to operate. That is the rock on which we base our lives.

Try to imagine what it would be like if your reality was different every time you woke up, (eg. if things constantly changed shape, or water flowed uphill, or your desk started to speak to you – ie. if the Laws of Physics no longer applied). What could ever give you a reference point from which to gauge what was real, where you were, and what to do?

With this in mind I hope you'll agree that our sense of a solid and stable environment is one of the most important and desirable aspects of our daily lives – even if we do take it for granted.

That yardstick is needed to make sense of all the other things that happen around us. Yet this comment, in itself, also points to the fact that a **totally** stable and fixed universe is neither what we observe, nor what we want. Without movement, and the change that such activity will bring, there could be no life, so we need change as much as we need stability. It's just that we don't want the level of change to become excessive. We don't want to lose our yardstick.

There are several explanations of how limited change comes about against a background of stability. The ones that are persuasive will fundamentally shape your perception of reality.

The second core factor in helping us to gauge the truth about our reality, (our other key yardstick), is the belief that if you can predict the future then your understanding of reality is probably correct.

The Laws of Physics have been very successful in regularly predicting what physical reality will do, and this is the reason why we believe in many scientific explanations. However it is also true that some scientific explanations have failed to predict the future, so in these areas it is 'back to the drawing board'.

The reason why I am emphasising this point is that while most scientific findings have sat comfortably within the mould of traditional thinking, others have been breaking the *principles* behind those views of the world in some fundamental areas.

Many of these 'contrary findings' arise in the deeper workings of nature, and they have spawned an entirely 'new' scientific discipline – Quantum Mechanics, as mentioned earlier. Yet over the past 40 years or so, a succession of remarkable discoveries has also raised challenges at the level of reality that we occupy, (not just the inner workings of atoms).

The facts are not in dispute and yet traditional thinking has failed to explain them. After a while it becomes necessary to at least consider other possibilities, and this is why people have begun to re-examine some philosophical principles and ideas. But where do we put our 'stake in the ground'?

In those many areas where there is still uncertainty a less robust type of yardstick can be gained by establishing the *range* of options about what the facts may represent. The range itself acts as a measure when we compare one thing to another.

Put another way, you have to be aware of the extremes before you can make sense of the middle ground, and while those outer edges may seem far-fetched, **they can be fun** even if you are not persuaded. More importantly they all carry significant ideas which you may find useful in other contexts... and they may help to explain the full diversity/richness of experience that we enjoy.

In summary therefore, the strange thing about uncertainty is that it increases the range of possible explanations... but in order to be credible each notion should still match the facts. Conversely, facts serve to narrow the range of possibilities; they rarely provide a full answer.

So new facts which shift our focus within the range of possibilities are very important.

We are in the strange position at the moment that some new facts are so challenging to common beliefs that while they are not hidden they are not being promoted and therefore they are failing to drive public debates in the media. They are so unexpected and so contrary to established thinking that science cannot form a view about them.

That caution would be fine if there were just one or two of these, but there have been a long list of them, and the realization is growing that traditional explanations may never be found because the issues they raise do contradict core principles.

Successful scientific theories are unlikely to change, but new findings may be shifting our perception of the overall context - possibly requiring an expansion of our thinking. The best thing to do is to have a look at the issues & ideas before making up our mind.

Moving on, when it comes to questions about the origin of existence philosophy is allowed to think in abstract terms when there is no human to observe and test something. The only evidence it requires is hard logic and it shows that there can only be **two** core options for prime origin – eternal existence or pure creation. Philosophy can work forwards from simple beginnings like this to establish some fundamental truths. It has also developed arguments which can be used to judge theories of origin.

On the other hand, science can only work backwards from the end results based on hard evidence plus an assumption that we know all of the factors involved. Because awareness of our complex reality is still growing, this cannot be a precise exercise.

Put another way, it is only after science forms an impression of what reality **is** that it can work in reverse sequence to see how it may have arisen, using patterns in the way that things operate – the Laws of Physics. Yet even these proven concepts depend on the assumption that we know all of the factors involved, and that is a big ask. Without that guarantee such exercises remain informed speculation even if they are very useful.

Those doubts are reinforced when we see that any backwards view based on cause & effect reaches a point where there are certain ***impossibilities*** that have to be overcome (Chapter 3). This is where religious ideas about God have found resilience in the modern age. The impossibilities are there and have to be explained somehow, and if not by God, then by something else... but what? As yet there are no firm answers.

Based on what we know about the natural environment we can speculate about various natural mechanisms which could potentially achieve the exact Universe we observe today, however we will find that the vast majority have one or more serious issues which prevent them from providing a complete answer. This gives us a measure of their **viability**.

It is very interesting to see the very small number of 'stand-out options' that have no serious concerns. The trouble is that I can't explain these to you in just a couple of paragraphs – you need to understand the logic behind them.

Once we have assessed their chances of success, we can then form our personal opinion about how **likely** it is that each of those viable options may have existed. In this respect we should recognise that as things stand ***none*** of the mechanisms have direct evidence to show they were real, so as with God, they are a matter of faith.

God may or may not be a reality but in this context such a question can be a distraction to the points we are trying to demonstrate. I prefer to say that the impossibilities and other key factors have to be overcome, and the different ways of doing so will each suggest that certain capabilities must exist within the Universe. It is your personal choice whether to attribute those capabilities to God or something more mundane but equally unproven.

In these ways I intend to honour my declared intention to cover all perspectives in relation to the prime origin of existence, while providing relevance to all viewpoints.

It may come as another surprise to realise that __all__ scientific and philosophical ideas about the nature of our existence claim that our experience of life is <u>different</u> to the underlying reality. **We are all living in an illusion to some degree**.

The range of ideas runs from quite shallow illusions (close to what we experience), to very deep and even total illusions, (which are theoretically possible, but hard to accept without evidence to show that our experiences are a delusion). However the middle ground offers us numerous possibilities that many will find easy to accept.

At one end of the scale, the scientific illusion is relatively shallow because it believes that physical existence is actually real and has true substance, even if it doesn't operate as we think it does in everyday terms. We could liken the experience to 'driving a car' rather than controlling every piston, wheel or other mechanical component. For instance, we think that :-

- things such as metal and china are solid when in fact the findings of science indicate that atoms are like a force field surrounding a lot of empty space with a tiny dot of 'substance' in the middle.
- objects have colour, when science says that they may be colourless because it is *light* which has the colour, and the colour itself may be just 'vibration'.
- some objects are perfectly smooth when science says that the shape and influences of atoms mean that there cannot be a 'frictionless' surface, etc.

(For those who need a refresh on some of these points I present a straightforward summary of school basics at the start of Chapter 2 to bring everyone up to speed).

At the other end of the range, philosophies can be quite dramatic with some even *denying* that physical things exist at all[6] and that everything is a mental construct – a deep and total illusion.

Once again - there are reasonable justifications for all of these ideas because they all have to match the evidence, even if they seem to contradict our lifetime perceptions about what reality is. They can also be quite interesting as 'The Matrix' film trilogy demonstrated when it brought one of these ideas to life… but with a fun twist.

It may therefore seem strange that a book which is largely focused on physical reality should begin by talking about illusion. Yet **the choice for you is to determine how deep the illusion might go**, before you can then form an opinion about where our existence came from. You will hopefully see why the latest scientific findings may point to explanations that are 'further along the range' than you might expect.

As part of this exploration you will probably find yourself performing mental gymnastics as you are asked to view things from a new angle. Doing so may at first seem strange, and possibly hard work if these ideas are not familiar to you, but I ask you to stay with them as there are good reasons why these notions retain their validity... and I will try to make them as entertaining as possible.

To illustrate the point, let's end with an old joke.

> Sherlock Holmes and Dr. Watson went on a camping trip. In the middle of the night Holmes awoke and nudged his faithful friend. "Watson, look up and tell me what you see."
>
> Watson replied, "I see millions and millions of stars."
>
> "What does that tell you?"
>
> Watson pondered for a minute. "Astronomically, it tells me that there are millions of galaxies and potentially billions of planets. Astrologically, I observe that Saturn is in Leo. Theologically, I can see that God is all powerful and that we are small and insignificant. Meteorologically, I suspect that we will have a beautiful day tomorrow. Why, what does it tell you?"
>
> Holmes sighed. "It tells me that someone has stolen our tent."

I hope you enjoy the journey.

2   Different Types of Stuff

A physicist is an atom's way of knowing about atoms. - George Wald

Every living person shares the wonderful range of tactile and colourful sensations that our environment provides. These reveal the characteristics which different physical materials possess and it's those characteristics which will probably tell us how to recognise the underlying stuff of reality when we eventually isolate it.

However, before that, we need to seek out the different components of physical reality, and then gauge whether something is a 'pure' factor within existence, or a product of other more basic things... and this is what science has been doing.

To start us on that same journey let's briefly recap on some of our school science so that everyone will be able to follow the logic of later arguments. In just short of 4 pages I will cover some of the properties of light, the structure of atoms, the nature of forces, and a basic notion of fields, however if you are confident in these things, and why they are relevant, then please skip to section 2.2.

2.1.1   A Straightforward Recap of School Basics

Let's begin with 3 factors that are different to the way that we perceive reality

- *the light from the sun is a mix of coloured light having all the shades of the rainbow, (as demonstrated by rainbows, and glass prisms splitting light).* It is only when the full range of this spectrum is present and in balanced proportions that our minds interpret it as the colour white, otherwise we see a tint from the dominant colours which are present.

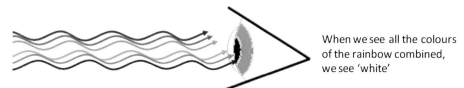

When we see all the colours of the rainbow combined, we see 'white'

- *Surfaces which we see as being smooth and even mirror-like are not smooth at all but are in fact very knobbly.* Scanning tunnelling microscopes allow us to see the outside of atoms, (but not inside them), and they show that each atom of an elementary chemical is the shape of a ball. Yet they can react with each other to form clumps/molecules, and each clump will have many lumps and bumps where the balls overlap. It's partly this lack of smoothness which provides the friction between objects that allows our fingers to get a grip on things. Yet, we all recognise that some surfaces are less knobbly than others.

What we might see as a smooth strand...

...is in fact a knobbly row of molecules generally made up of differently-sized atoms

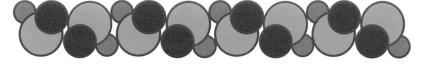

15

- *the atoms in our environment probably do not have any colour built-in.* The thing that gives each object the impression of having a colour is the ability of atoms to absorb some of the coloured light that hits them while reflecting the rest. Our eyes will only detect the colours which are <u>not</u> absorbed: we imagine that the object has the colour of the light which has bounced off it and which has been deflected into our eye.

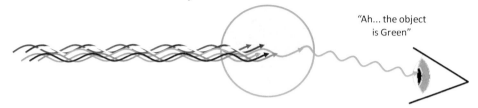

"Ah... the object is Green"

These explanations seem to deny our basic sense of what is happening in reality, suggesting that our Minds are presenting us with a distorted impression of what is actually occurring. Yet we seem to enjoy that illusion as it forms our entire lifetime's experience on Earth, and it also seems to allow us to see the bigger picture of what is happening rather than getting bogged-down in the mechanics.

Life would be very boring if all we observed were vibrating colourless atoms and different wavelengths of light being absorbed. So 'illusion' per-se isn't necessarily a bad thing. Indeed, to this extent it may be a blessing rather than something sinister.

Yet the illusion goes deeper when we try to interpret what is happening *within* atoms.

From our early chemistry lessons most of us will have seen that when we heat up solid materials, such as ice or metal, to a high enough temperature they turn into a liquid (eg. water) and then into a gas (eg. steam). We learn that heat causes the atoms and molecules which make up these substances to move further apart, so they increasingly lose their structure and 'firmness' to become liquids then gasses, however those atomic components do not actually become something different unless they undergo another chemical reaction internally.

Science originally suspected this because when these substances cool they return to their original more solid state, so they aren't fundamentally changed by this process. Modern microscopes and other technologies have now proven the case beyond doubt. We <u>can</u> see the outer 'shell' of atoms and how they are arranged in objects.

There are a wide variety of materials in the world and science explains this by saying that there are different types of atom. Atoms are not the base level of existence. The various types arise because they are constructed from different combinations of 26 types of smaller component inside them which *do* seem to be universal. Science believes this because it has been smashing atoms apart for many years and the same types of component always emerge from the destruction. We <u>cannot</u> see these things but we can detect their presence through the impacts/effect that they have on other objects, or the radiation that they leak like a vapour trail.

This type of analysis has revealed another illusion. At our higher level of existence, the level of atoms or bigger, material such as slabs of wood, plastic, rock, metal, or other substances, will appear to be solid. Yet science can also show that the tiny *component* particles of an atom, can be fired straight through sheets of solid metal etc. as if nothing was there at all – except for the odd dramatic particle that goes off at a wild angle. Both sets of evidence are true, so how can we reconcile the ideas?

In talking about his experiment and the scattering of alpha particles as he fired a stream of them through gold foil, Ernest Rutherford said, "It was as if you fired a 15-inch shell at a sheet of tissue paper and it came back to hit you."

Science has formed a theory that *the atoms which we and our environment are made of are **not solid** but are primarily made of near-empty space, with only a miniscule core that has any real substance.* It's necessary to delve into this if we're to understand some of the arguments later.

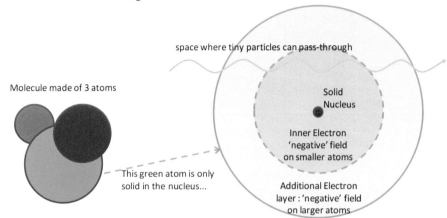

According to this theory, it's only when a sub-atomic particle is fired through an atom and it hits the core (or 'nucleus') that we register the occasional deflection. The nucleus contains almost all of the weight/mass of an atom but because it is such a tiny part, (proportionately much smaller than shown in the diagram above), most sub-atomic particles will pass straight through an atom.

Although this *is* a theory because we can't actually see what is happening dynamically **within** an atom, (even with today's largest microscopes and detectors), there is more and more evidence available to support this layered view of existence.

While atoms are not believed to be solid they <u>are</u> perceived to have a visible boundary even though this isn't a shell. (This is how we can see them). Science has often portrayed this boundary as the outer point of the orbit where one or more very tiny electrons spin around the nucleus very quickly, but more recently the electrons are perceived to spread out as a sort of 'fog' in a layer around the nucleus that generates a real 'force field' which defines the outer boundary.

Interestingly the 'heavier' an atom is the larger it tends to be, but this has little to do with the *size* of the nucleus, which is always very small. It primarily relates to the space that electrons occupy, which in turn relates to how many electrons an atom needs to maintain its stability.

An atomic nucleus is 'positively charged' while electrons are 'negatively charged', and although opposite charges tend to attract each other, within an atom (for one reason or another) they don't merge. Typically, the larger the core of an atom the stronger the positive charge that it will contain, and therefore the more electrons will be required to balance it. So if a talkative atom were ever to claim that it had lost an electron you could be sure that it was positive about it!

Science also finds that there are limits to the number of electrons which can co-exist in the same space and state around the nucleus, so if more than this number of electrons is required within an atom, the extra electrons will create new layers of space to occupy, around the original layer, (as shown in the diagram).

Various scientific experiments reveal the layered structure of atoms and their component parts. These have ultimately provided quite a complex explanation for why they act as they do.

Moving on, science has identified another set of distinctive factors in the physical world: the 4 basic forces in nature, which either hold things in place to create structure, or generate movement. To set a context, atoms have structure, as do the objects which are formed by collections of atoms. But what holds those things in place? The 'Strong' and 'Weak' forces only operate within an atom, and can largely be thought of as 'holding' forces which create that internal structure.

The 3rd force of 'Electromagnetism' exists both within and outside an atom, underpinning all chemical reactions, which provide many larger scale structures as well as movement.

> Once objects are in motion they can themselves exert a force if they collide with something else, however they don't represent a prime source of movement – the factor which gets things going in the first place.

Science has a degree of understanding about how these 3 forces achieve their effects, (which we will consider in Chapter 5), yet there is a continuing mystery about the 4th force – Gravity. We see its effects across vast distances and note that it primarily operates outside atoms, however at this stage science hasn't established **why** it works.

The forces of electromagnetism and gravity are **prime sources of movement** for groups of atoms, (objects). If there is direct contact, (one thing crashing into another), it's easy to see why a still object might start to move. But if there isn't a collision then some other prime factor must generate movement, and this arises when :-

- we see objects expand or shrink as they get hotter or colder, (eg. metal bars, or tiles on a roof, taking up more space and pushing other things out of the way).
- the space atoms occupy changes when they either join together/merge, or the chemical reactions split the molecules into smaller groups.
- the atoms can mysteriously attract or repel each other.

The real mystery is *how* these two forces can pull or push things across empty space. We see that a magnet draws metal objects towards it even though there is no direct contact, while gravity seems to work on any material and across very large distances – even in the emptiness of space. In these circumstances there is no hidden string connecting the objects to cause them to move together, (like winding in a reel), and the influence can still be present even if we pass our fingers between the two objects.

As kids many of us will have scattered iron filings over a piece paper which is lying on top of a magnet in order to reveal the field of influence which exists around it. So what could a field like this be? I have heard some people liken it to a hidden fluid, (such as a mass of water), where the physical 'substance' in that liquid deflects/pulls/pushes another object as it passes-through the water.

However in the emptiness of space it becomes all too apparent that deflections cannot explain the type of effect that we observe. These characteristics of fields distinguish them as being something distinct in nature and may perhaps point us towards the underlying stuff of existence.

So while we know a lot about the effects that forces and fields of influence achieve, we don't have a clear idea about what they *are* or *how* they do what they do.

2.2 Candidates for the Core Stuff of Reality

A key question at the heart of our debate is : What is our physical reality made of?

From the previous section I think we will all realise that the answer will not be easy to derive, so let's ask a more simple question that may take us towards our goal.

How many different *types of 'stuff'* lie at the core of our physical reality?

In asking this question we are looking for **base factors** of existence, which means that we have to find ways to distinguish them from the myriad of other items which are not 'core'. As we don't find them neatly labelled as relevant or not, we have to identify them by their distinctive characteristics and the principles of how they operate. But what are these features?

It has taken thousands of years for mankind to learn what is relevant and what is not... and even then we cannot be sure that we have identified all of the right things until we understand the totality. All I can do is to present you with options based on the knowledge we have gained so far.

Many of you will have heard the ancient theory that everything is made of 4 underlying elements – Earth, Air, Fire, and Water. The development of scientific understanding has now shown this to be incorrect, because we now know that Earth, Air and Water are all made of atoms, which in turn are made of common sub-components, so the underlying stuff of existence must lie at a deeper level. Yet Fire seems to be different to the other three, representing fluid energy and power rather than rigid physical substance, so in recent centuries, many believed that the obvious differences between them pointed to two base factors of existence.

However thanks to Albert Einstein[4] and his famous equation $E = MC^2$ the modern view is that there may only be <u>one</u> type of stuff underpinning physical reality. That's because the equation suggests that mass and energy are interchangeable – and many people will therefore conclude that they are made of the same stuff, (typically energy).

I am only aware of one theory which attempts to explain how fluid energy might gain form & structure, and that is highly speculative. We will consider it in chapter 5, yet something like this will ultimately be necessary if people continue to believe that mass and energy are made of the same underlying substance. Because uncertainty remains around this question, and many people are puzzled that mass and energy demonstrate very different characteristics when they are supposedly made of the same stuff, people try to think how Einstein's equation could be interpreted in other ways.

Such ideas are also highly speculative, but some of them are based on established theories and have therefore gained some acceptance. As an example:-

Mass has both structure and substance/'weight' – neither of which are characteristics that we associate with raw energy. Yet we know that when mass is destroyed it releases vast amounts of energy. Could it be that a 'unit' of mass is like a container which stores a huge set amount of energy? If the container is then broken it would release a fixed amount of energy in proportions demonstrated by Einstein's formula.

How such containers might be formed is also considered in Chapter 5.

As a result of Einstein's equation, a philosophy has grown that there is only one type of stuff which underpins <u>everything</u> in reality - the underlying substance of ***physical*** existence. This is known as '**Materialism**' and it's a reasonable philosophy if you accept that *everything* shares the same set of characteristics and principles, (although there are many who think that they don't). Yet if you accept the materialist viewpoint there are implications which may ***not*** seem entirely desirable – even if they turn out to be true. Let me explain…

Science has traditionally said that the balance between stability and change is achieved because physical existence operates strictly on the basis of 'cause & effect', (sometimes referred-to as 'Causality' and the philosophy of 'Determinism'). This operating principle is held out to be a fundamental characteristic of physical matter normally adopted by Materialism.

More specifically scientists have argued that a ***precise*** set of opening conditions can only lead to **one** specific effect, and therefore the pattern of change is controlled. You **cannot** have several possible outcomes, and **nothing new** can arise to 'break the mould' or introduce change because the rules <u>have</u> to be followed. In other words, events can only unfold in one way. If we originally perceive an event to have more than one possible outcome, then closer examination will ultimately reveal that something must be subtly different in the originating circumstances for each outcome.

Yet there are implications from this logic. If correct... everything that ever happens, everything you ever think about, and everything you ever do will <u>always</u> be inevitable.

As an example, Materialism cannot deny Thought but it has to explain it through physical matter and always using the principle of single cause: single effect.

> Existence will be merely acting-out a pre-set script. You will **not** have Free Will and you could not take full responsibility for your actions.
>
> Praising geniuses or punishing criminals would seem wrong because they had no choice in what they thought or what they did – they were acting out a fixed script.
>
> This underpins the philosophy of 'Materialism' based in strict causality.

This may sound strange to many of you, but we should remember that things may still be inevitable even if we can't predict them, (because we may not know all of the factors at play). Materialists **can** present rational ways to make their theory of inevitability work in most circumstances, but not all.

Partly for this reason, not everyone agrees with that explanation. More importantly it doesn't fit-in with their experience of life or what they want from it - including the legal/political factor that we <u>do</u> want people to take responsibility for their actions.

To enable Free Will, true creativity, and true change, (rather than just an inevitable outcome), it's suggested that one or more ***additional factors*** must exist alongside physical reality to provide that greater level of flexibility, allowing existence to operate as we commonly perceive it – but not to the extent that change gets out of control.

In other words, there may be other types of 'stuff' that underpin existence.

If correct, this alternate stuff would have to demonstrate fundamentally different or opposing characteristics to physical matter/energy to show that it cannot come from that source. A key example would be if we could break the principle of cause and effect, but this would be equally true of any other factor that went against the key characteristics of physical matter as we know it.

As headline examples of this from 'recent' scientific findings (which we will see later):

- communication and influence can happen **instantly** over large distances, (at least 10,000 times the speed of light[19]), thereby breaking a fundamental Law of Physics.
- the 'Dual Slit' experiments (Ch.6) have suggested that some sub-atomic particles may be entirely without shape/form until they are observed or measured[29].
- 'Paired Particles' which have no physical link and can be miles apart, will actively copy/mirror changes in each other's state of being. (Ch.7) Crude awareness?
- the Quantum Eraser experiments (Ch.6) suggest that a surviving paired particle may be able to retrospectively change information about its twin after it has died.

These things directly contradict the traditional explanations of Newton and Einstein, and have left their supporters reeling. When you fully understand what they mean these findings are likely to shift your view of reality as well – at least to some degree.

Chapters 6 & 7 will present the facts and then explore the potential for real world explanations - allowing you to make a choice. Yet, all of the current proposals seem to have profound implications. Where could they take us next?

These results were **not** what I expected to find when I began my research for this book. However, for the time being I want to stay focussed on the principle of cause & effect because I want to set a framework for our ongoing debate. If anything can challenge the principle of Causality then it will establish a polar opposite perspective to Materialism, and set the boundaries for our philosophical range of possibilities.

Assessing some examples will help you to form your own opinion.

2.2.1 Randomness, Spontaneity, & Thought

It may come as a surprise to many who haven't studied traditional science, that it doesn't do 'Random' or 'Spontaneous'. As we have seen, the core principle which it advocates is that every action requires a prior cause, and that a precise set of opening circumstances can only lead to <u>one</u> specific effect. If similar scenarios appear to produce *different* outcomes/effects it's because

- the starting position must have been different, and/or
- additional factors were present to make each situation unique.

The scientific principle of 'cause & effect' is a fundamental requirement of physical Matter/Energy. (I should point out that many scientists today will talk loosely about the 'non-determinism' or lack of apparent causality at the sub-atomic level of physical matter – but that is not the same meaning implied by philosophy[8a] – as we shall see).

The opposites of cause and effect are spontaneity and randomness, so anything which displays these properties <u>cannot</u> be based in physical Matter/Energy if 'physical causality' is correct. It's unfortunate that there aren't more specific terms available, as the words I've used can get applied very loosely, so in making this point I need to be quite precise about what I mean:

- Spontaneous - something arises or occurs without a cause, but which might be logical, structured, or intuitive.
- Random - there can be more than one outcome from a precise start point - outcomes can vary for no reason.

In other words I am referring to factors that can break the principle of cause & effect.

Some of you will no doubt be sceptical about what I have said about the scientific view of how things work, so I feel I ought to explain a little more.

We must distinguish the words 'randomness' and 'spontaneity' from the adjective 'unpredictable'. Traditional science believes that physical events can have outcomes which we **cannot yet predict**, but ultimately, if you knew all of the facts, you would see that they are inevitable. It says that <u>everything</u> physical can be explained by very precise factors. For example:-

> Lottery machines try to mimic random activity, but are actually just *unpredictable* in their outcomes. If we choose the example of stirring-up balls in a drum, it's very difficult to monitor what is happening exactly, but if we did, science could tell you precisely why we got each lottery result as each ball was pushed with different strengths from different angles.

> Truly *spontaneous* and *random* behaviour would be if the lottery balls were not stirred and one just jumped up and selected itself. Another example would be if one pool ball was struck by another at a specific angle, but instead of just being able to head for one pocket it could potentially head towards any pocket.

Of course we don't see random/spontaneous acts like this occurring in physical matter but we **can** potentially see it occurring in Thought. If you put someone in the middle of a circular room with ten identical doors, and then asked them to move to the exit...

- their decision to move 'now' rather than 'then' would be spontaneous, (without an inevitable prompt at an inevitable time), and

- their choice of door would be random – regardless of the knowledge they brought with them. (To counter such things as a 'straight ahead start point' you could put them on a rotating platform, etc., to ensure randomness).

A motorised ball in that room would simply follow pre-coded instructions immediately.

Determinism says that a new idea must always have a prior cause, (you must always work up towards it), whereas spontaneity says that it can just pop into your head.

> *A truly new thought or decision is a moment of creation that <u>science</u> cannot explain, because it requires either randomness or spontaneity.*

We can also illustrate the differences through our use of computers which are programmed to operate in <u>fixed</u> ways, (even if they are set-up to identify different circumstances requiring different pre-determined responses).

If you imagine using your pc plus the fixed instructions which represent your word processing software, we can see that computers don't produce just one set text from this fixed routine. The range of written material coming off the printers is enormously varied because the critical factor which changes the outcomes is the human user, who puts different things into that fixed process. Computers are not creative or spontaneous; people are... unless the rigid principle of causality is true and we are all acting-out a fixed script while living in an illusion.

If we **are** being genuinely creative then Thought could point us to a fundamentally different type of underlying 'stuff' compared to physical Matter/Energy. We don't know what that stuff is, but it would have the characteristics of randomness & spontaneity which could also enable Thought to occur. (When some scientists controversially suggest that randomness has been demonstrated at the sub-atomic level of existence, they are supporting attributes which have distinguished Thought from physical Matter/Energy). I am trying to illustrate the case for other types of stuff.

Constant references to 'underlying stuff' become very tedious and potentially confusing if there is more than one type, so a 'shorthand' has developed which can be slightly misleading. In common parlance people will simply talk of 'Matter/Energy' as opposed to 'Thought' as short-cuts for the underlying stuff that they represent. In one sense these mark the extremes of opinion, but there are other possibilities which are less easy to label. We will come to these in a moment.

Until the middle of the last century the success of scientific analysis into the workings of atoms and the materials/objects we perceive in our daily lives, led many to believe that spontaneity and randomness were myths, and that Thought, (as a distinct type of stuff), was a myth as well. This was partly a reaction to the abuses which religious extremists had inflicted on men of science for centuries, (eg. through the Inquisition), and so there has been a tendency for Materialists to rail against anything which even vaguely supports the notion of God operating through His thoughts/will.

It also reflects the fact that we still can't point to Thought and say 'here it is'. Yet to suggest that electrical signals in the brain are the essence of Thought is incorrect. Electricity doesn't think and nor do neurons as far as we can tell. As Book 2 will explain, the logic & evidence point to deeper factors at work.

Hopefully today we can move beyond prejudice against God to consider that Thought may represent something other than the workings of Matter/Energy. I'm not saying that it **is** based on other stuff, just arguing that we should consider the possibilities when Materialist explanations are falling short. This is true of other things as well.

Ultimately science is here to reveal the truth and not to reinforce prejudice from either side. Yet if there is truth in religion it must, in the end, be squared with science.

The modern justification for considering other types of stuff comes from scientific experiments that are explained in Chapters 4, 6 & 7.

I am **not** talking here about telekinesis or other psychic phenomena. I am talking about the real workings of Matter/Energy, which are relatively simple to understand even if some of the details provoke some mental gymnastics. It's not dull and they really are worth a bit of effort to read even if you're not scientific.

So in overview, to frame our view of the world, Materialism & Determinism will essentially argue that everything comes from Matter/Energy which in turn is based on cause and effect, and no thought will ever be truly creative or random. From the inspiration of Einstein to the music of Mozart or the Beatles, it would all be inevitable.

Put another way, the Materialists/Determinist view of the world has to demonstrate that nothing new can arise without a cause, and that everything works **within** the rules, (Laws of Physics), and therefore cannot change them.

> The grey area, (and a prime reason why Materialism remains distinct from Determinism), is that within the realm of Quantum Mechanics, some people now argue that there may be a ***lack of determinism*** at the sub-atomic level of physical matter, (ie. a presence of spontaneity and randomness), although this is highly disputed for reasons we will come onto shortly.

> If this were true, it would break the traditional explanations of Materialism and validate the attributes which have been denied so often in the past. It would

also give added credence to ideas currently presented as 'different types of stuff'. For this reason the suggestion is strongly resisted by many scientists.

People who *disagree* with Materialism have to demonstrate the opposite - that either:

- the rules **can** be changed/broken;   or
- that something without cause can be added into the mix;   or
- that a precise start point can have more than one possible outcome.

So let's follow the alternate line of belief concerning the nature of Thought.

Our perception of reality begins in our head. We look out on the world from inside our head and we interpret everything we learn about our existence through our Minds. In order to build a reliable perception of the truth behind our Minds, we have to find aspects of our existence that are unquestionably true.

René Descartes[5] provided us with one of these foundations for belief when he said

"I think, therefore I am."

To be clear, he was saying that we **can** be sure that our minds are real, but we need other things to demonstrate the reality of physical substance. (To avoid big distractions let me just say that there are **no** *conclusive* ways to do this, so 'Idealists' who believe that all of our reality is in the Mind still have a viable case – see below).

If physical substance is real then Descartes' truth still tells us that any consideration of existence must explain the Mind as much as anything else, because it shapes our perception of reality.

Thought is real. We use it every day and via our bodies it generates physical effects, yet it seems to show quite different characteristics to the core properties of physical matter. In headline terms many people feel that **it doesn't always operate on the basis of 'cause and effect'**. Thought appears to start new things from scratch and can also influence physical matter. Think about it…

> Thought may be the **only** thing that can make physical matter deviate from its inevitable chemical path… even if we can't yet rationalise how it manages to be random or spontaneous.   Buildings don't construct themselves...
>
> Can you think of any such deviations that occur without Thought?

But is this just an illusion? Could Thought still be part of an inevitable Materialist mechanism? We will consider some examples to test these ideas in section 2.3 below.

If you conclude that Thought **does** point to a fundamentally different type of stuff then we have to consider its influence in other aspects of our reality. That could be huge or minimal depending on both your point of view and the evidence available.

Alternatively, you may either keep an open mind, or conclude that there **isn't** another type of stuff, in which case you have to address the scientific findings which seem to defy explanation by the normal rules of existence, (for instance, the results from the Dual Slit and Bell's Theorem tests – chapters 6&7).

Some people argue that there are aspects of the **material** world we are not yet aware of, which provide additional capabilities; however we would then have to demonstrate why those characteristics are so separate from the normal workings of Matter/Energy. To my mind that separation means there is little practical difference to ideas about other types of stuff, which are easier concepts.

Thinking 'outside the box' will always be challenging, but it is also true that in many scenarios, hard logic keeps pushing us towards the **necessary presence of spontaneity or randomness** to explain how prime origin and fundamental change might occur, (eg, if existence had a beginning it would need a spontaneous act to get it going).

If physical matter/energy can only explain the next step in an inevitable sequence it can't be the source of any true/fundamental change, so something else would need to be, and there is currently only one credible candidate – the underlying stuff of Thought.

Of course we don't know what the future may yet reveal, so we must also consider the middle ground - other types of stuff which may not be polar opposites to Matter/Energy, but permutations on a theme. Science is already beginning to formally acknowledge and explore such ideas in order to explain some findings in cosmology. It uses terms such as 'Dark Matter' and 'Dark Energy' to plug gaps in theory, yet the more we consider what these substances are required to do, the more they seem to require special properties – representing a third type of stuff?

Again – don't worry about the terminology, I will explain later.

For the moment I just want to complete the outline framework for our debate.

2.2.2   The Philosophical Range

If Materialism represents a *physical* perception of what reality is, the other end of the scale is '**Idealism**', which in its extreme form suggests that everything is in the Mind and that physical matter is entirely an illusion[6]. Put another way, Idealism says we are living in a dream world, where the imaginary environment we construct for ourselves always follows the same mental rules, giving us our sense of structure.

Before any of you scoff at this too much, it came as a great surprise to me that there is now experimental scientific evidence that may support key aspects of this perspective, (see Chapters 6 & 7). Again – this is nothing to do with psychic activity.

Idealism is a philosophy that also says there is <u>just one</u> type of underlying 'stuff' – but it **isn't** physical matter: it is Intangible Thought... whatever that stuff might be.

This may be hard for most people to accept but in toned-down circumstances, the 'Matrix' film trilogy showed how our entire sense of reality could be a trick of the mind without any substance. What we *experience* would be real, but the circumstances would be a deep illusion.

Both Materialism and Idealism are said to be 'Monist' philosophies, (only having one underlying element to existence).

Yet it's natural to think ..."Why can't we have both?" This would give us the stability of physical existence and the spontaneity/creativity of Thought to introduce change. Perhaps Life can only come when these two core factors work together?

Theories which suggest that there are two types of stuff underpinning existence are known as 'Dualist' philosophies, while those which suggest 'two or more' are categorised as being 'Pluralist'. This gives us a range of possibility.

The different permutations form a smooth transition between Materialism and Idealism and many people will find their personal beliefs lie somewhere in the middle ground. Evidence can shape your beliefs, but equally your beliefs can shape the way you interpret the evidence, so you need to understand the key aspects of each argument to make up your own mind.

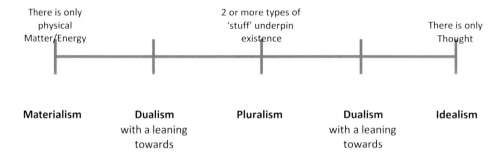

There is only physical Matter/Energy		2 or more types of 'stuff' underpin existence		There is only Thought
Materialism	**Dualism** with a leaning towards **Materialism**	**Pluralism**	**Dualism** with a leaning towards **Idealism**	**Idealism**

To me, it was a surprise to find that most of the dominant theories have only one or two core factors underpinning existence. This is because credibility demands that you indicate what those 'elements' might be. Physical Matter and Thought seem to be the prime options as well as the extremes of the range, because they have properties which are polar opposites. Time/Movement may be another, (you can't have one without the other), so there are notions which refer to three specific elements.

Beyond this, suggestions of 4 or more factors tend to become entirely theoretical and less specific about what those other underlying factors might be, or how they manifest themselves, yet the effects they seek to explain are very real. The proven ability to communicate and influence things faster than the speed of light (Ch.7) may point to other dimensions, which would have to contain very different types of stuff.

I have already referred to 'Dark Energy' which is a concept being pursued by NASA to explain astronomical findings that I will explain later. The Double Slit experiments have defied Materialist explanations but some of the key characteristics directly point to properties of other stuff which may actually tie back to 'Dark Energy' (see Ch. 6).

As all of these perspectives apply their own layers of illusion, while providing *viable* explanations for our reality, we inevitably ask ourselves, 'What can we believe'?

This is where we return to our base level yardstick - the rigid consistency of our everyday experiences, which will lead most people to conclude that physical reality does have true substance – agreeing with the core scientific starting position. I feel it's genuinely significant that everyone begins their life with this instinctive position, which is also the basis on which we live our daily lives.

Another key challenge to the Idealist view of existence is that doctors/scientists have investigated our brains and discovered that their complex tissues are the source of some control in our bodies, (although not all – see Book 2). Idealists cannot deny this link between our Minds and our 'physical' brains and so they basically have to argue that brains are a necessary part of the illusion.

Other than it being counter to our everyday experiences the main objection to the Idealist perspective was that it lacked any evidence. However that has begun to change with the results from the Double Slit & Bell's Theorem experiments (Ch.6&7). These mean that Idealism cannot simply be dismissed although it will seem highly unlikely to many people. Even if correct, Idealism doesn't have to be the full story.

As an example, if our Minds control our sense of reality, and the physical world didn't actually exist, it shouldn't be able to surprise us by invading our space. But it does.

I was recently standing in my kitchen making a coffee, and was shocked when I turned to grab the sugar and banged my head on the cooker hood.

At that moment my mind wasn't planning it to be there, yet circumstances didn't bend to accommodate my expectations. It gave me a great sense of independent physical reality, (rather than illusion)... as well as a lump.

Dualists on the other hand can tap into the strengths of both of the main philosophies. Let me illustrate how their philosophy might work in practise.

Their concept of Thought is that it is based in a 2nd type of underlying stuff which has an independent existence from Matter/Energy. Our minds would effectively be making decisions **outside** the physical tissues of the brain. The purpose of a physical brain would therefore be to convert Intangible Thought into physical actions. So our Minds would merely be using the brain as a way to :-

- generate physical effects through our bodies,
- receive information back from the physical universe, and
- store information in the physical environment.

In this scenario both types of underlying stuff would be enhanced by working together. Physical life would be a 'collaboration' between them, meaning that it would have the ability to be random or spontaneous while benefitting from the stability of physical matter, including a body which had real substance.

In contrast, the philosophy of Materialism/Determinism concludes that our thoughts are purely based on physical components, say electricity (electrons) plus the physical 'wiring' of both our brain and our central nervous system. As the philosopher Daniel Dennet once famously wrote :

> "The trouble with brains is that when you look in them,
> you discover that there's nobody home."[7]

[Dualists often counter this by saying that the destructive examination of brain tissue will inevitably break the link between an intangible mind and physical Matter/Energy, so you're unlikely to find any Thought lingering there.]

I think we've done enough to illustrate the different lines of thinking, but the important thing to remember is that they all have a viable fit with the facts.

So if your personal philosophy says that you're a creative person with Free Will then I thank you for choosing to read my book. On the other hand if you believe that thoughts are physical and mechanical then I recognise that you had no choice in doing so, but I still hope that you enjoy it.

Before we leave this topic, and to complete the explanation for the range of thinking, we must consider whether the different factors underpinning reality are true *sources* of existence or whether *some may have emerged from things which **are** source factors*.

That may sound a bit dull but it's important for us all to have an idea of what these notions mean in practise. So please stay with it. I promise to be brief, but there will be a few more mental gymnastics along the way!

Materialism and Idealism are polarised views which both claim to identify a single factor that underpins existence, so they must also represent fundamental 'sources' of existence which must have been present for as long as there was reality. On the same logic, the position occupied by the label 'Pluralism' says that two or more factors have <u>always</u> co-existed.

Yet these notions do not sit in splendid isolation. They form part of a range with many intermediate ideas between them generating a seamless spread of opinion. You can't have two and a half types of stuff underpinning reality, so we need to consider what some these intermediate views might represent.

In short, they either portray mixtures of the true base factors, or they represent the possibility that one type of stuff enabled another type of stuff to emerge.

The top line of the diagram below provides an outline of how each concept might work, and you will see how the ideas transition from 'single underlying sources of existence' on the ends, to multiple co-existing sources in the middle.

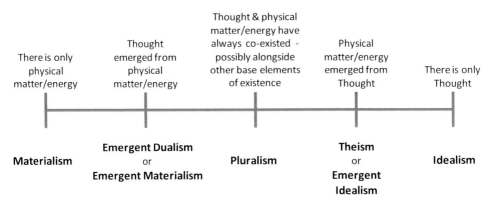

'**Emergent Materialism**', on the left hand side, typically suggests that Thought emerged from Physical Matter/Energy to become a distinct factor in existence… but not a source factor.

Conversely, on the right hand side, *Theism* suggests that Thought led to the emergence of physical matter, and it is here, that we find traditional concepts of God which position Him as 'the creator' whose will or Thought brought physical matter into existence. There <u>are</u> non-God theories within this space but I'm not aware of an official term for such views, however you might call them '**Emergent Idealism**'.

Atheism is a denial of God, but it doesn't have to deny Thought, so atheistic views can occupy any position along the range. This is equally true of *religious beliefs which do not include God* but which <u>do</u> include the soul, (eg. Buddhism and Jainism).

Conversely, and to round off this line of thinking, the only way for God-based theories to cover the full range of concepts, from Materialism to Idealism, would be to alter your definition of God…. and there have been various definitions of God across the centuries. Yet it is not necessary for the notion of God to cover all possibilities within the range. Perhaps being specific to a particular position within that range could even be a strength. Having said this, Atheists do find robustness in covering all angles because they suggest that their view will always be the answer. Everyone else will inevitably disagree!

In terms of the diagram there is one more concept we need to consider, and, it could be seen as adding a third dimension to our line.

'*Property Dualism*' is a twist which says that there is only one type of underlying stuff, but that it has two sets of properties which are revealed when it is either placed in different circumstances, or is accessed in different ways. One of those sets of properties relates to physical substance, while the other relates to Thought and Mind.

In religious terms this can be likened to the ancient Taoist philosophy which says that there is one source of infinite potential (The Tao) that constantly sub-divides into ever-more specialist areas, including life and physical matter to give us the full range of different materials, as well as the other aspects of existence.

Until science fully accepts randomness & spontaneity, or can fully explain the origin and nature of existence without them, it will face a constant series of challenges about life and how the properties of Thought might be achieved.

For this reason some scientists argue that while Newtonian Physics, (eg. cause and effect), has been shown to have limitations, the new generation of scientific principles based around Quantum Mechanics[8] does give us 'Random' effects – but it doesn't[8a].

As science cannot prove the existence of a second type of stuff it tends to adopt a Materialist perspective and through Quantum Mechanics it has simply found a way to deal with *unpredictability* where we can't see what's happening.

In other words it is still looking for missing factors to explain differing results within the principles of cause & effect. The fact that there are different outcomes from similar experiments doesn't **prove** a lack of determinism – it may just mean that we can't see the missing factor. To support this view it can be pointed-out that the range of outcomes from such experiments is limited, with most outcomes being repeated many times. So there's still a pattern of behaviour that implies rules/causes.

Yet it is equally true that the activities of sub-atomic particles *could* be providing the first direct evidence for randomness and spontaneity. We can't see what's happening.

Quantum Mechanics describes the *range of possibilities* for any scenario which it cannot observe directly, (and such ranges could theoretically be very large indeed). How scientists narrow the options in order to develop rules which can make useful predictions about what will happen becomes very important, as we'll see later.

For the moment, let's just say that Quantum Mechanics doesn't explain things in the way that laymen would expect. It really gives us the likelihood of different potential outcomes, ie. a set of probabilities about what would happen mechanically if we only knew more, and these possible outcomes are almost always based on logical causes and effects... although they may occasionally incorporate observed results which have no known causal explanation... yet.

2.3   Is there an Absence of Cause and Effect in Decision-Making?

In terms of framing the range of opinion about different types of stuff I earlier marked one set of boundaries/distinguishing features by the presence or absence of randomness and spontaneity in the universe which cannot arise through Matter/Energy.

If you **can't** identify a substance with these capabilities, there needs to be some reasoned justification for believing that it exists when you search for it, and if this stuff is one of only two or three things that underpin existence, we ought to see plenty of indications that it is indeed working in the environment. To validate the range of possibilities we need some specific examples to prove a point one way or the other.

As Quantum Mechanics can't prove whether these abilities exist, we have to look elsewhere. I have therefore been looking for examples where Thought can do things which Matter/Energy is deemed unable to do via cause and effect.

If even a single example can be found then it will break the principle of causality and potentially establish Thought as something very different in its own right.

The answer we find will guide our personal view of everything that exists and how it may have arisen… so we must consider if there are any examples of decisions which demonstrate spontaneity or randomness – the opposites of cause and effect.

Yet that's not as easy as it may sound. In order to consider the arguments we first have to narrow the focus of the issues.

Let's assume for the moment that our Thoughts can indeed generate a new idea from scratch without a cause. In most cases we will be layering a new idea on top of an existing pool of background knowledge, a history of preferences, and some personal objectives to guide our choices. So it would be wrong to assume that any notion we might generate is *completely* original because we never start from a 'zero' position.

> Let's say that I suddenly decide to go for a walk through the town. My pool of knowledge has allowed me to know that there is a town and that I have the ability to control my body sufficiently to move away from my present location. It has probably given me an outline of the possible routes I might take. I also know enough to interpret and navigate around the environment that I will encounter, even if it is unfamiliar.

> If I was a highly sophisticated robot in a world entirely governed by cause and effect then everything would be pre-determined from the start. My decision to travel would be a response to a need for me to travel, and I would be pre-set to respond in an inevitable way to the circumstances that I would encounter en- route. To all intents and purposes the same journey might be undertaken whether or not I was a 'spontaneous' being. So where might the difference lie? What would make my thoughts different to such a robot?

> If I am truly a spontaneous and free thinking being then I need to add or change something within the base position, and in this case the clearest demonstration of that ability would be in the original decision to undertake the journey in the first place. If there was no need for me to travel but I decided to do so anyway, where did that decision come from? Was it truly creative and spontaneous (ie. without cause), or was it an inevitable outcome from a complex series of circumstances that included my memories, the way I felt at that moment, and the things that I could sense from my surroundings?

Put another way, with 'strict cause and effect' the pot of circumstances doesn't change and events play out in an inevitable way against pre-set rules. The question of 'What set the rules' is open to debate later, but for the moment, if we are trying to demonstrate spontaneity, we have to distinguish the things that **change** or get **added** to either our knowledge or the background circumstances, in order to create a new scenario, no matter how small that may be.

In contrast, Materialists need to demonstrate that circumstances were following an inevitable pattern of cause & effect where pre-set causes would explain every situation, even if the people involved were not consciously aware of those factors.

A useful starting point is to consider how 'making a choice' from a range of similar options can lead to an <u>inevitable</u> outcome. Here the challenge for Materialists is to demonstrate that a free choice wasn't random. The standard argument used is that we will always rigorously apply a ***ranking*** to the options available and we will always choose the item at the top of the list, because it will reflect our best logical preference – making it an inevitable/unavoidable selection against the priorities of the moment.

> As an example, when faced with a range of drinks to choose from, we will assess how we feel, and then refer to our memories of what the different drinks tasted like to determine which is most likely to satisfy our preferences, needs, or desires of the moment – a ranking.

While this notion might seem perfectly reasonable we each have to decide whether it can be applied in every circumstance that we can think of, and whether the ***mechanism*** of prioritising things can be adequately supported by <u>identifiable</u> needs, preferences or desires that fit the moment.

Unfortunately, as you get deeper into this debate, you will see that examples will arise where no obvious factors can be thought-of to rank the options, and it is often the case that in these circumstances Materialists will resort to comments such as 'a myriad of unknown influences' to explain a decision. You must ask yourself in each case whether that is a satisfactory explanation, or whether this points to an example of randomness or spontaneity.

In this battle-ground I think it's very hard to prove a case **unless you are able to strip away most, if not all, of the other potential influences on us**. To this end I have found that simple 'negative' examples are the most effective in demonstrating possible examples of spontaneity. That may not mean much to you now but a series of examples may illustrate what I'm getting at. The challenge is on!

If you rank things, it would be fair to assume that you'd do so in 'order of preference' and therefore desirability. Yet it's an undisputed fact that we don't always choose the nicest options, (indeed, it's not unusual for people to select something which they actively dislike).

Such decisions can be done for logical reasons because something which we deem to be more important takes priority, as in the case of unblocking a drain. Yet in this case there's little alternative. We need to find examples where there are clear choices and yet people still take the counter-intuitive option – illustrating why there may be more than a simple ranking when we decide on something. Unfortunately the strongest examples are the starkest and they take us into uncomfortable territory. Apologies.

Let's start with the example of people choosing to have potentially painful or unpleasant forms of sex. I **am** talking about free choices here, not rape or other forced circumstances. So how could that be an inevitable free choice – top of the list?

Well there are many potential reasons, for instance :

- Wanting to please their partner whom they love.
- Curiosity.
- Wanting to prove that they can do it.
- They may like the idea, or even get something positive from it in some way.

As these examples show, in this and many similar circumstances Materialists can present numerous 'side issues/factors' that could take priority over the potential discomfort to justify the person's logical decision to accept an unpleasant experience. Only the individuals themselves will be able to judge whether those guessed reasons were correct, yet these factors do generally make random choices less believable.

Do 'ultimate choices' fare any better in demonstrating the ***potential*** for spontaneity or counter-intuitive decision-making?

> Deliberate self sacrifice on the battlefield is perhaps a stronger example. Such acts can be even more difficult to explain where the dead soldier has saved somebody he was known to actively dislike and whom he could have allowed to be killed instead.

> What about the sniper who sees the order to withdraw but decides to stay in an area being overrun by the enemy, from which there is no real prospect of escape later?

> We can speculate about: 'purely following routine'; or 'an overwhelming hatred of the enemy'; or standing up for a principle; simply making a mistake; or a desire to end it all. But we shouldn't dismiss the possibility that individuals may simply make a different choice which may be 'random'.

Even so, these examples still leave some room to doubt that the decision was spontaneous rather than inevitable. We need to look elsewhere.

The simplest examples can often be the strongest ones to establish evidence for an effect, and they may also show that if spontaneity or randomness are indeed possible then they may not be rare events at all. By way of example :-

> I may, for no reason, suddenly decide to move my arm:
> * raising it higher, or lowering it
> * waving it about,
> * touching my nose, etc.

> The choice that I make doesn't have to be driven by a reaction to external events. I may simply decide to do it in a particular way, for no particular reason.

Yet even these possibilities might have a secret cause – for instance a subtle ache that my body senses but my conscious mind doesn't really register, causing me to flex my arm in a certain way. We cannot prove that our precise movement wasn't driven by an inevitable factor, but equally the materialist explanation is now highly speculative.

We have to find examples which seem to purely originate within our minds and are not driven by a response to a cause. This is why I like negative decision-making. By this, I generally mean the decision to stop doing something rather than to start it. So...

> How long is a pause in conversation?

> Well, it's as long as you care to make it, because under neutral circumstances there's no rule that would make it an inevitable period of time. It is also a decision that's entirely under your control, so you could make it just that bit longer or shorter. Indeed you could potentially make it last for hours or even forever.

So in a situation without external pressure, or moments of forgetfulness, the decision about when to end a pause either has to be random or spontaneous. A **deliberate** pause may even be the best example, because you judge how long it will be. It comes from within your Mind.

Try it with a friend, and determine if your choice is inevitable or not.

If Materialists/Determinists talk about everything being an inevitable pre-determined outcome then here we may see an example which **does seem to break the mould** – making it a strong argument for the Dualist and Idealist positions – that Thought is fundamentally different to physical Matter/Energy because it doesn't have to strictly follow the rules, and can be random or spontaneous.

The natural world and chemistry are not capable of having a pause, but thinking beings can, and if a simple basic choice shows clear signs of spontaneity or randomness, then it establishes the principle that can be applied through all of the other examples I presented earlier, and more.

It remains a personal choice whether you accept these examples or not, however once you make your choice you have to follow the consequences/implications from that philosophy. So what are some of those implications?

Because of the two principles of

i)   'cause and effect', and

ii)   'nothing ever disappears, it only gets transformed into something else',

science is able to encode its Laws and theories within mathematical equations: where a perfect balance has to be achieved, (for instance between prescribed inputs on one side and outputs from a process on the other side). For the most part, it's considered that the entire physical environment works to balance such equations and cannot step outside the rules. *So what could write new equations, (new Laws of Physics)* when this is a necessary part of almost all theories of origin based on an eternal existence?

Inevitable processes should be put forward to explain **all** changes – but they aren't.

Cause & effect doesn't allow for 'mistakes', so logic suggests only 2 possibilities:

- external factors - but in this context it would probably have to be external to all Matter/Energy in our 3D universe. Other universes? God?
- spontaneous or random influences – ie. other types of stuff.

If we are purely physical beings then the full range of human creative achievements would also have to be explained in inevitable ways - including our lack of Free Will, and as we've seen this has not been demonstrated.

On the other hand, if people do consider the possibility of spontaneous influences to explain change and creativity, they should also consider how that influence could be prevented from turning everything into chaos. The way in which we use Thought may give us the most natural way to explore such possibilities.

An obvious argument in Dualism is that chaos is avoided by only having a narrow touch point between the two factors, and that touch point is structured to provide a strong level of control. This narrow touch point would probably be the brain.

Which of these descriptions do you feel most closely match your own experiences?

2.4 Where does this leave us?

Whether or not these exercises persuaded you that there is true spontaneity or randomness in the world, the examples should hopefully explain why there are different viewpoints, which probably cannot be dismissed even if you are not persuaded by some of them. We have established a range of possibilities.

The jury is still out and you will inevitably have your strong preferences.

There is no doubt that in recent centuries science has leant strongly towards Materialism, (possibly because we haven't been able to isolate and test Thought). Equally, society has leant towards Dualism because we believe that individuals make decisions that are not inevitable and therefore we want them to be creative as well as responsible for their actions.

You could say that society has been living with a split personality. ☺

What I and many scientists have found utterly surprising is that experiments which were intended to virtually kill off any opposition to Materialism have ended up supporting other views. This isn't just a case of having to find a 'missing factor' to bridge a gap, these facts seem to fundamentally challenge the core principles on which Materialism and Determinism are based.

We will be looking at some of these things in more detail over the coming chapters to help you make up your own mind.

Yet for me, it is the number and consistent pattern of these findings that give added credence to some of the other perspectives. They are no longer just wishful thinking.

At the same time, the proven ability of scientific rules to predict what will happen in the physical environment means that those principles will remain true even if we eventually set them in an Idealist context. But at the moment there are a growing number of scientific findings which are pointing us to something else as well.

In some respects, talk of a 2nd type of underlying substance is a metaphor for saying that certain characteristics exist which are distinct from the substance we know of as Matter/Energy. That separation is a key factor because it has set people's perceptions about what existence is made of.

If these characteristics are real it doesn't really matter whether they are ultimately explained as the influence of other types of stuff, other dimensions, or deeper layers of existence, (ie. factors within material existence), they would amount to the same thing – separate capabilities, long denied. All of the theory relating to other types of stuff could then be applied through these other mechanisms, forming the basis of true controlled change and possibly an explanation of Thought.

This will certainly colour our perspectives when we look at potential explanations of prime origin, which is a subject that has some very distinct characteristics. Yet if you don't accept the existence of any capabilities that go beyond the familiar properties of Matter/Energy you have to explain the findings with a much more restricted 'armoury'.

The case for Materialism is far from dead, however it now has to re-think its approach and come up with new answers. The scale of that task will become apparent as you read on, however I will show you how some lateral thinking and the use of philosophy can provide very straightforward **Materialist** solutions to some of the deeply challenging findings from science which have defied explanation for many decades, such as the Dual Slit experiment, (Chapter 6). But there will be other explanations too.

My reasons for doing this will be threefold: these ideas may

- point to some solutions that tell us more about the nature of existence
- help us to see what mechanisms could possibly produce the spread and patterns of movement that we observe in galaxies across the cosmos.
- look back in time to explore some viable options for the origin of existence.

By identifying the key findings which challenge our perceptions of reality we hopefully get to the core issues which will ultimately unravel the mystery. These will also provide us with a practical tool, in the form of a checklist, with which to assess all ideas about origin.

You will then be able to re-shape your own opinions about physical reality, and who knows, we may even start to bridge the gap between religious and scientific thinking.

3   Basic Views of Origin

Before you can form an opinion about the prime origin of everything you need a way to look back into the past. In addition, you need to consider what it is that you're looking for. If we were looking for the origin of an egg then we would go looking for things that can produce eggs. But the fundamental/ultimate/prime origin of existence is quite different as we don't really know what the stuff of existence actually is.

The logic of causality, (where events can only unfold in one way), gives us a basis for looking back into the past to some degree, once we understand the rigid principles by which nature operates, and once we are convinced that we know all of the factors at play. However in truth we can't say that we know either of these things with any precision in relation to our reverse engineering.

To make the situation worse, if any spontaneous or random events have occurred, the backwards path is likely to be severely blurred or broken at those points in time because they'll no longer be able to suggest what the previous steps might have been.

The only hope for our understanding of early existence is if the opportunities to exercise spontaneity or randomness are confined to very limited areas/circumstances. This may actually be the case because we don't observe a chaotic environment, and transforming events for the Universe would seem to be rare.

Equally, on current evidence and on a much smaller scale, living beings may be the only *regular* source of such influence. If this is correct then the general story of how our planet evolved to this point would be almost completely **un**affected by our activities, (we arrived too late and our influence is too small). On similar logic we could also assume that the impact of **all** physical beings across the cosmos would be negligible compared to the scale of the Universe.

So without large scale randomness & spontaneity, there **are** reasonable grounds to speculate about the main events of the past, (as long as we don't expect precision).

Centuries of scientific & philosophical analysis also tells us that our sense of reality begins inside our heads, where our minds form an impression of what is around us, based on the signals we receive from the wider world. We know that our minds are real.

Yet 'impressions' are not physical things and, as we know, the philosophy of Idealism says that our Universe may just be a trick of the Mind. So in basic terms **we still have to prove that physical substance is real** - even if this is the working assumption behind everything we do in our day to day lives.

The reasons why we trust those signals and believe in the real substance of the physical world, boil down to 3 main factors :-

i)   our physical environment is always present in our conscious state.

ii)   the mechanics of our Universe operate in <u>exactly the same way</u> every time we use them, (ie. on the basis of fixed rules).

iii) the wider world influences us even more than we influence it, (ie. something or someone else must exist beyond our individual selves).

Put another way, it's the rigid consistency of our experiences with the wider world (something **ex**ternal to ourselves), that leads us to believe the physical Universe has true substance, and I feel it's significant that in terms of day to day life we <u>all</u> approach it in this way, (even if theory kicks-in later to tell us that it's an illusion).

The belief in causality allows scientists to encode our understanding of how things work within mathematical formulae, where one side of the equation has to <u>exactly</u> balance the other side. It doesn't tell us what enforces those strict rules of operation, or what put them in place, but to the extent that these equations always correctly *predict* an effect from a cause, we trust them.

An equation is often held out as proof of a theory if it correctly predicts the future. Yet it remains theory rather than absolute fact, because there have been many discarded equations across time. However in broad terms the most reliable theories become the Laws of Physics, and in doing so they enshrine the principle of causality on which they are based.

> It is only in the more recent studies of Quantum Mechanics, (which investigate a lower, more fundamental level of reality), that evidence has begun to mount about a more fluid and less structured form of our existence – as we will see later in Chapters 5-7.

Equations simply encode an idea within a 'mechanism' of cause & effect, yet they must always be compared to reality, because if they don't match they can't be right.

The relevance of equations to this debate and the reason why I am mentioning them now is because certain ideas about the origin of existence have been entirely derived from manipulating equations rather than looking at the reality they try to explain.

We can't go back into the past to check that such views match reality and worse still we can't know all of the factors that were involved in events millions of years ago. (We don't even know all of the factors at play in the current moment)!

As a result, there are dangers for scientists who form theories about reality which are entirely based on manipulating equations. Those equations will <u>inevitably</u> replay the principles and assumptions on which they are based, whether they're correct or not.

To the extent that a set of ideas fits-in with other successful equations, there is added confidence that these might be successful too, but in the broader picture **equations test viability; they are not proof in themselves**. Only 'undisputed facts' about reality can act as proof – one way or the other.

In this respect it's important to remember that **all** evidence about our physical reality comes from the brief period of history in which modern man has existed.

Put more simply, all evidence gathered by human beings is drawn from the **current moment**, and if we draw conclusions from it about the past, that is our interpretation.

Humanity wasn't present to witness **any** events in the lifetime of the Universe, other than in the last 'blink of an eye'. We may therefore *assume* that a rock we have just dug up is old, or that the light we have just received has travelled 13 billion years across the Universe to reach us in this moment, but that is 'interpretation' even if there are good grounds for believing it.

Each statement has to be moderated by a level of confidence, so in the above examples you may have more confidence that the rock is old than in the implied age of some light. Yet in order to form such judgements you have to understand the facts and their logical interpretation which led to the declarations.

Facts serve to reduce the number of possible explanations, but they rarely leave you with just one option. Equally, theories of origin need to <u>exactly</u> explain all facts within the reality of today.

3.1 Can there be a true beginning?

This debate about origin can become very clouded by interpretations of evidence, so I think it is important to keep the basics in mind at all times. In terms of logical possibilities things either had a beginning or they didn't.

For many people the instant reaction to this comment is that talk of having 'no beginning' is just silly. Everything must have a beginning – it's a fact of life that we see every day. In simple terms, if something is here today that wasn't here before, then it must have had a beginning.

Yet the logic of prime origin is quite strange and can lead us to other conclusions – which stem from the scientific principle of causality.

We tend to think that beginnings are like the start of a line – something stark and clear, yet that may not always be the case. For instance, in the spectrum of light we gradually see red turning to orange then yellow then green then blue etc... and it becomes very hard to see where a beginning might be.

In a similar way science points us towards a continuing *process* of change with many factors at play, that argues against all start points. Yet what put that process in place?

No matter how far back we take things the logic of cause and effect will always need to have a sub-component or previous stage of evolution that requires an earlier step. The rationale cannot stop unless we suppose that something started without a cause – ie. something spontaneous occurred that can act as a Prime/Ultimate beginning.

Put another way, the dilemma at the heart of this scenario is that ;
if there was an ultimate beginning
anything that triggered it cannot have had a beginning.

In many cultures God has traditionally provided the prime way to reconcile this conflict of impossibilities, and it's very likely that this is where the modern notion originated, (unless you believe that God planted the awareness of Him in us). Most of us will be familiar with the 'Creator God' explanation, and it's probably a good time to re-state it briefly here:

In the beginning the thoughts of God, through the power of His will,
created the physical Universe which conformed to His design.

In the West, the desire for a physical beginning including the spontaneous triggering of physical creation, (first cause), is perhaps the strongest popular argument for God which is supported by the predominant religions. It is commonly seen as a 'proof' that's based on our observations of how things are, plus a logic which seems instinctive to most people. It is a theory that also begins in the Mind... the mind of God.

The logic for a Creator God requires Him to either be eternal, or deems that He must have emerged spontaneously, (without cause). Across the ages, different religions have adopted one or other view. However the additional need for Him/Her/Them to 'create' or 'design' things, has instinctively led us to position God within Thought, turning the concept of God from a 'mere' reconciling factor into a being.

In philosophy the logic of 'first cause' is commonly known as the *Cosmological Argument* and its origins are often said to have emerged from a group of Islamic writers from the 9[9]th century to the middle-ages who belonged to the Kalam[9] school of philosophy - so it can also be known as the Kalamic Cosmological Argument.

However God is not the only way to explain existence. If something has to either emerge spontaneously or have no beginning, why shouldn't that be Matter/Energy itself? This is the core of atheistic thinking.

Because Matter/Energy is not generally believed to be capable of creating itself spontaneously, I find that most Materialists/Atheists will opt for physical Matter/Energy being eternal. This is supported by one interesting observation from science – that the 'stock' of existence always seems to remain the same size. Nothing within our current physical reality ever seems to be completely destroyed – it is always re-cycled. For instance, if you burn something it turns into the same amount of 'stuff' but now it takes the form of smoke, ash and radiated heat energy. Nothing is physically lost, it is 'just' transformed, all-be-it in a rather unpleasant way.

Yet the practical difficulty with arguing that there was no physical beginning is that *science itself tells us there was a start point for our universe in a huge explosion – the Big Bang, 13.77 billion years ago*, (see Chapter 4).

The evidence on which this is based is another reality check. It strongly suggests that there was a *single* beginning, not multiple start points, but we have to ask whether this event represents spontaneous 'pure-creation' - out of absolutely nothing?

A pure creation event without God or Thought, would have to mean that a complete absence of anything spontaneously created physical reality, (without a cause), yet everything we know about the natural physical environment tells us that it cannot do spontaneous things.

There are other difficulties with this notion, too. Science believes that a Universe could exist in many viable forms, (different Laws of Physics), yet *very very few* of these would be able to support life as we know it. As our Universe does support life, an explanation based on an initial burst of spontaneous pure creation would also have to assume that existence bucked serious odds to **get it right first time**.

So if we assumed that under very special primordial circumstances, (unknown), physical matter might indeed be able to spontaneously create itself in a single event, this explanation would be relying on **luck** to make it viable – and that is not generally considered acceptable where a specific outcome is required... ie. our exact Universe.

In this context the only way to overcome the 'luck' factor would be to suggest that nature had many attempts at creation – trying-out every possibility, and eventually hitting on our version of reality... however science has never observed any characteristics of our Universe which may point to a number of creation events. The current evidence is for just one event – anything else is pure speculation.

This is another reason why Materialism has tended to promote the idea that physical existence has **always existed** and that an eternal 'cause and effect' process has allowed nature to experiment with different forms of Universe, until it achieved our version of reality. The eternal process is most likely to represent a **cycle** of activity that ultimately leads to a major event which either transforms the Universe or creates a new one with a different configuration. The logic for one such mechanism, The Bang-Crunch cycle, is explained in Ch 4. and this is not based on spontaneous pure creation.

So let's try to summarise the thinking so far.

The idea of no beginning is an impossibility to some, but it is an inevitable conclusion within the rigid principle of causality. Yet it can seem equally impossible to suggest that something can happen without a cause – like pure creation out of nothing. One of these supposed impossibilities must in fact be possible. But which?

We have to look for evidence from what is real, and some things <u>do</u> have a clear beginning and they do seem to require a spontaneous act.

We normally identify a drawn line, buildings, cars, planes, mobile phones, etc., as having beginnings, however these are all things that **we** initiate, and we may be one of the few sources of spontaneity in the Universe, (ie. our minds may be a source of spontaneity, so each of our actions might represent small new beginnings).

The alternate Materialist suggestion is that living Thought is simply one inevitable effect within the natural process of physical change which underpins the environment, (ie. it is within causality). So even the creation of a piece of music would be **inevitable**.

We looked at the examples supporting both sides of this argument in Chapter 2.

This is where we come back to basics and personal choices. If you believe that existence is **incapable** of being spontaneous, then it cannot have a beginning. Alternatively if you feel that true beginnings are a reality, (the basis of the Cosmological Argument), then you must either believe in a source of spontaneity, or an external factor, (such as God), exerting its influence.

In the midst of this, the Big Bang is a reality check, but we have to carefully consider what it might mean. Chapter 4 will do this, and we will see how recent scientific findings are perhaps swinging the logic back in favour of spontaneity.

3.1.1 Other Principles about Prime Origin

I heard a tale about one unfortunate man who was hauled before the Inquisition because he claimed that the secret of life and existence lay in flour. He had reached his conclusion because 'everyone' had seen that if you left a clean bag of flour alone for a few months, living weevils would miraculously appear within it.

In terms of mediaeval science the basic observation appeared to be true, yet the explanation may have been lacking a few factors. In terms of our awareness today, the conclusion ignored the fact that the flour, (and possibly something else), had to exist already. Without scientific evidence, theories of origin can be many and varied.

Instead of looking for celestial bags of flour in the night sky, science has looked for the most fundamental elements of our environment within atoms, and has found that these tiny objects are made up of electrons plus protons and neutrons which in their turn are made up of hadrons, quarks etc.

It suggests that each of these may have an earlier origin in energy and force because Einstein[4] (through his formula $E=MC^2$), has successfully shown that mass & energy are interchangeable. Yet energy is something - so where did that come from?

With the earlier concepts in mind there are only 3 generic ways in which the Universe might come to exist : either

  a) The underlying stuff of reality has always existed without beginning, or
  b) Existence spontaneously emerged, without prior cause and indeed, without anything existing beforehand, or
  c) Spontaneous creation occurred **within** an eternal existence, (a combination of the above), which includes the traditional explanation of a Creator God .

Hopefully we can all see that beliefs about origin must split along these lines.

The viability of each of these options hinges on the presence or absence of spontaneity or randomness. In this respect, if society recognises a true ability for us to choose our own actions in life, (ie. that events are not inevitable), then it must also recognise the presence of spontaneity, opening-up options 'b' & 'c' above.

It is only the 'Materialist/Determinist' form of atheism which argues for rigid cause & effect, and a pre-set existence where we are just acting-out a script.

So we reach another set of personal decisions :-

1. Has anything always existed, (without beginning), and if so, what?
2. Has there ever been pure creation, (adding to what was there before), and if so, what?
3. Does pure creation have to be triggered by something that already existed, or is it possible to have spontaneous **self** creation out of absolutely nothing?

The Cosmological Argument comes from an assumption that all material things have to be <u>created</u> – marking a fundamental change to what was there before, and therefore a start point. While we have just explored the logic for physical creation, there are other ideas about what a first fundamental change might have been.

For example, the theory put forward by the Greek philosophers Aristotle and to a degree Plato[10] suggested that the underlying stuff of physical matter had always existed **but had been static**, and as a result they were more focussed on the '<u>first movement</u>' that triggered all other movement, and which ultimately fashioned a transformation.

If something has existed alone for an eternity without movement, then we can see why the generation of the first movement would still count as a spontaneous act, (ie. without cause). In a similar vein we can say that if any process has lasted for all eternity it must be in a stable condition that would never change by itself.

From this we can derive an important principle which will influence our later interpretation of the evidence. In a closed system with no **ex**ternal factor:

Any **change** to an **eternal** pattern of existence requires an element of spontaneity

Against many expectations, it is this logical conclusion which undermines traditional Materialist explanations of origin based on the Big Bang, when we combine it with recent scientific discoveries/facts that have refined our sanity check, (see Ch. 4).

Even if Matter/Energy is believed to have existed without beginning, we will find as our analysis deepens, that there's an ever increasing need for spontaneity/randomness to have occurred at some point to explain some of the changes we perceive in the past.

Many people will turn to the **underlying stuff** of Thought as the missing factor because it is the only suspected source of spontaneity and randomness that we have. If you believe in the scenario at all, it's your choice whether to believe this was the Thought of God, or an unthinking act by a 'different type of stuff'.

The level of spontaneous change we have been talking about would require Thought to influence physical matter, and there's some justification for such thinking, if you consider that our minds do control our bodies.

So if you eventually conclude that spontaneity **is** necessary to explain the general state of existence today, we can therefore see ways to rationalise it through Thought.

On the other hand, if we suppose that Intangible Thought was the only stuff to originally exist, we have to work a lot harder to explain how physical matter might be generated when it didn't exist before.

I am only aware of three generic ways to rationalise how this might be achieved, based on what we imagine the underlying stuff of existence to be. (There may be others, but for now it's important to show that some explanations are possible).

1. Intangible stuff may be eternal but physical Matter/Energy requires Pure Creation to bring it into existence.

2. The underlying stuff of existence is eternal and can be distilled/split into two or more distinct elements with unique characteristics – eg. Matter/Energy vs the stuff of Life & Thought. This doesn't have to be an even split. Outputs could range from a lot of Thought + little Energy, to a 3-way split of balanced opposites, (per Taoism)

3. Only Intangible Stuff can be generated by spontaneous pure creation, but it can be converted into physical substance through a common factor, (such as Energy).

We will explore such ideas in later chapters, not to claim a solution to the origin of existence, but to show that viable ways can be found to justify different theories.

For the moment, let us just say that collectively, these notions mark out the range of viable possibilities, so let's look again at the diagram which I presented in Chapter 2.

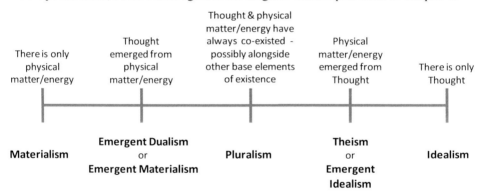

As indicated, these philosophies tend to suppose that 'the pot of existence' has either stayed the same, or that things have been **added** to the pot through pure creation.

It is very rare for people to suggest that anything has disappeared from existence and that the pot has shrunk, although this is a logical possibility. (As we have seen, science says that when physical matter is destroyed it is converted back into its base element - energy - and that nothing is actually lost).

Pure Creation, (something <u>out of absolutely nothing</u>), is a nice idea and we can all imagine things suddenly appearing as if by magic. The movies have even presented such notions on screen, however this is like witnessing the results without suggesting how it could be done, which is why many people, and some religions, still consider it to be an impossible act for physical matter. Yet when we consider the possible emergence of Thought (ie. ideas just popping into our heads), we can see the potential wiggle room for spontaneous creation.

Again - in terms of fundamental origin, **the only alternative to pure creation** is the *Transformation/configuration* of something that already exists.

Mainstream science doesn't have an official position on how the Big Bang may have been triggered or whether there was a prior existence of any sort. Its focus has been on generating a model of how events may have unfolded **after** the explosion began.

The fundamental problem for a mathematical description of pure creation is that the equation couldn't balance. There would be nothing beforehand, but something afterwards. Mathematically therefore there can only be eternal existence without beginning – which is why the evidence for the Big Bang and the latest findings concerning the expansion of the Universe (Chapter 4) are such a problem.

Yet the enduring appeal of a beginning, (which implies that something came out of nothing), has led people to develop 3 other approaches to the origin of existence which can also 'balance the equation':-

a) The first is *Induction*, (explained later in section 5.2.2) which can be thought of as moving energy from one place to another.

b) The second way is to invent 'negative matter' in order to balance the positive matter that we experience, so that the 'creation equation' balances to zero.

c) The final main approach is to ***re-define 'nothingness'*** in ways that can make it something, thus turning creation into transformation.

As we develop our thinking, it will be important to remember that the concepts of creation, transformation, & induction all require different underlying capabilities within existence. We will explore theories of origin more closely in Chapters 4, 9 & 10.

There are two final sets of ideas which will influence the basis of any concept of origin. Firstly, the objective isn't simply to find ways to get things going. Explanations of existence have to bring us to the reality that we observe and experience now… in every respect.

Working forwards from any deemed starting point, using the mechanisms which might explain how things came to be, we have to achieve our reality without resorting to luck in any aspect of existence. We ***can't*** simply assume that the right conditions just happened to occur. Put another way, our Universe doesn't just need mechanisms to exist: they must also operate with a particular combination of 'strength settings' and possibly in a very specific sequence.

There are a myriad of possible settings, and any fledgling universe with a serious imbalance could easily tear itself apart and not survive. Yet there are many configurations which science believes could be stable/viable, although only a tiny proportion of these might be able to support life as we know it.

Any explanation of existence therefore has to explain how these **settings** came about with such precision, co-ordination, & uniformity - overcoming unbelievably huge odds.

The last principle within the debate is that *complex* objects can only emerge in 3 ways:

a) Through spontaneous creation of that complex object, (ie. it just popped into existence fully formed – requiring spontaneity),   or

b) Through a process of evolution where a series of events *based on cause and effect* transformed a more simple reality <u>in an inevitable way</u>,   or

c) Through a process of evolution ***that is not inevitable***, and therefore either influenced by design, (which is likely to be based in Thought), or through spontaneous or random acts.

Only option 'b' gives us an explanation which doesn't involve spontaneity or randomness, and even then we would have to ask 'What determined the start point'? Was that inevitable too, or did it also require spontaneity or randomness?

3.1.2   Impossibilities & Dilemmas about Existence

Sooner or later every generation confronts the ultimate questions of origin & existence, and each comes to realise the difficulties lying within the principles of the debate. The elements of each argument often contradict each other to leave everyone perplexed about what to believe. As we've seen, we cannot even be sure that an ultimate beginning is possible at all.

When faced with some of the 'Impossibilities' at the core of the 'origin debate', it becomes harder and harder to explain physical reality without a reliance on luck, or the presence of a second underlying stuff of existence with other capabilities.

For instance, let's look at how the issue of explaining change can lead us in very different directions.

Regardless of whether physical existence is eternal or had a starting point, if we say that it rigidly operates on the basis of 'cause and effect' and that everything which happens can only unfold in one way, the course of this inevitable path must in turn relate back to the way in which existence was set up. This is particularly true if the inevitable path cannot cycle through all permutations. If the range of outcomes that could be achieved is limited, or could just provide one flavour of reality, we would be reliant on luck or design to achieve that initial set up.

The only way to avoid luck or design would be if every permutation of existence lay on the inevitable course that was to follow, which must include a mechanism for **fundamental** change. Yet how can fundamental change happen without spontaneity; chance/randomness; or an **ex**ternal factor?

In this respect Materialism has a real problem because spontaneity and randomness are deemed to be beyond the capabilities of Matter/Energy, and **ex**ternal factors imply that something very different must exist. It is interesting how often Materialism just assumes that such change is possible, when this is the critical point about origin.

> For instance there is a dilemma because any eternal cycle of change must be based on a mechanism that is preserved, (otherwise it couldn't be expected to operate for all eternity), and yet fundamental change would seem likely to break that mechanism.

> So we do need Materialists to explain how the two aspects can operate together – which they haven't really done for any of the proposed mechanisms.

> It's not acceptable just to assume that the exactly the same stuff will somehow produce radically different outcomes. That is randomness, (which they deny), unless there is a specific explanation.

As we have seen, the core of this debate only has 3 ideas about our prime origin.:-
pure creation,   eternal existence,   or a combination of the two.

Determining which of the 3 you believe in, rests on how you overcome the *6 perceived impossibilities* which I mentioned that lie at the heart of this debate.

All 6 relate to perceived capabilities of the underlying stuff.

The Impossibilities

Either on their own or in combination, these impossibilities always leave us with the dilemma that some of them must in fact be possible. Consider any doctrine concerning an ultimate origin and you'll find that at some point it will either ask you to prefer some impossibilities over others, or suggest that in some precise circumstances some of the impossibilities might be avoided. It is perhaps the mix of impossibilities which distinguishes one doctrine from another. So what are they?

It's time to list the 6 that I have identified. It is impossible:-

i) to conceive how anything could have no beginning.

ii) to rationalise **how** something/anything could come from absolutely nothing.

iii) to guarantee one particular outcome out of many possibilities unless the single evolutionary path is set to explore **every** possibility within the time available.

iv) to have a prime/ultimate beginning, or a change to an eternal pattern of events, without a spontaneous/random act or an **ex**ternal factor.

v) for physical/mechanical things to perform a spontaneous/random act.

vi) to rationalise **how** spontaneous or random events might occur.

Because these impossibilities have formed such an impenetrable barrier to logical explanation, over time there has been a subtle grading of impossibility – ie. that some ideas may be less impossible or more believable than others.

After all, we do exist - so something has to break through the logical barriers.

An obvious start point is to recognise that just because we can't yet explain something doesn't mean it's impossible. We need a reality check to see if there is any evidence to show that some of these impossibilities do in fact occur.

The *logical* evidence for causality provides grounds for believing that option 'i' may be possible, (something may have no beginning), while the evidence of Chapter 2.2 shows there are reasons to believe that spontaneous or random events *can* occur, so there are also grounds to believe that option 'vi' may be possible.

We can further illustrate the 'wiggle-room' by contrasting 'ii' & 'i' above:

- we might just about accept the possibility of something having no beginning, because we cannot naturally envisage a beginning or end to Time either

whereas for most people ...

- the suggestion that something can come from absolutely nothing is totally impossible, because a complete absence of anything can't do anything. Yet we have to ask whether the absence of anything tangible is the same as an absence of an ability or the potential to do things?

So it depends what you believe to be possible – with all of the associated implications for each philosophy.

It is worth you pausing here to test your own ideas of creation and see how they match-up against the impossibilities. Also as we explore the range of ideas later, you might wish to do the same comparison on each new concept to see where it struggles.

4   The Big Bang Theory

In case anyone doesn't know the Big Bang theory let me describe it. Science suggests that all matter/energy in the Universe exploded from a point in space some 13.77 billion years ago[11]. This figure was verified by a NASA probe (WMAP[1]) launched in 2001, and has been adopted as the best figure we have to date.

Supporters of this theory generally believe that all of the physical matter/energy in the entire Universe was compressed at that first moment, within a pinprick location in space known as a 'singularity', and it has been expanding outwards in all directions ever since, creating the ball shape of the universe that we popularly imagine. If it is correct the theory still leaves open the underlying question - whether this was a moment of pure creation, or of re-cycling/transformation?

It's hard to conceive that all matter/energy from the entire universe could fit into such a small place, but the original concept from science believed that the force of gravity generated by all of the physical matter in existence, would have become so strong in a singularity that it would remove all of the empty space that exists within each atom, (a vast proportion), so that everything would fit tightly into that one point.

For the purposes of this debate it's irrelevant whether the degree of compression resulted in a golf ball, or indeed a planet, rather than a pinhead. The main issue is that science says that all matter/energy from the entire Universe was present in that singularity, (ie. nothing was created later). In addition, this material was in an extreme state, (the raw underlying stuff of physical matter), where none of the current 'rules' of existence were present.

The singularity concept is important if you feel that existence was transformed or re-shaped and not created from nothing at that moment. Yet if you feel that material <u>was</u> created out of nothing then of course the process of creation might produce a stream of existence rather than everything being in the same tiny place at the same moment.

Science doesn't have an official view about what existed immediately before the Big Bang, however if there had been a prior Universe its look and feel probably wouldn't have borne any resemblance to the Universe we know today as the fabric of existence would have begun afresh at the Big Bang, (ie. being assembled using different rules).

Religious communities which believe in a creator God will often imply that this explosion is the point in time where physical reality was either created by God out of nothing, or the earlier 'chaos' was given shape by Him as part of a structured design. Yet we can see that these two options play to both sides of the possible explanations.

The important factor for them is that the Big Bang theory seems to mark a ***beginning,*** requiring spontaneity or randomness, appearing to vindicate what many mainstream religions have argued for centuries, yet religions don't necessarily declare this officially as history has taught them to be cautious about what science may uncover next.

The alternate interpretation is that the last Big Bang event was merely the latest in a long line of such events, and it therefore didn't represent a creation event, but merely a continuing, unthinking process which was also based on 'cause & effect'.

Although the Big Bang concept emerged from the implications of Einstein's Theory of Relativity, the main evidence for it came from the observation of galaxies across the Universe which demonstrated that virtually <u>all</u> galaxies were moving ***away*** from us, (only one is coming towards us), which suggested an explosion.

We'll consider whether this is the only interpretation of the evidence in later chapters.

We refer to the theory by its populist name[12] which emerged from a slightly disparaging remark made in debates during the 1940s about whether this notion of an expanding Universe was better than its rival, the 'Steady State Theory', (which essentially suggested that the Universe had remained broadly the same for all time).

While the Big Bang idea reinforces some religious concepts of a creation event, the alternative 'Steady State' theory also had supporters from both the religious and atheistic communities. The Steady State theory was favoured by many atheists who wanted to show that matter had existed without beginning, and was also supported by religious believers who suggested that God had designed all of the sophisticated elements of the universe and then put them fully formed into their particular places around the cosmos, 'willing all things into being' as we see them now, (keeping us dependent on God and maintaining an alignment with some aspects of religious texts).

Different forms of the Steady State theory were proposed to fit various scientific findings, but it became increasingly impractical. However for people such as Stephen Hawking, it was evidence about the smoothness of background radiation in the Universe that put the 'final nail in the coffin', as scattered galaxies sitting in a fixed position could not produce the even levels of radiation that we do record. (The radiation given-off by galaxies in fixed positions should appear to be stronger as we looked towards them and weaker or non-existent elsewhere – but it wasn't).

To some extent therefore, it may be argued that a key difference in approach between these two notions was that the Steady State Theory was 'concept seeking evidence', whereas the 'Big Bang' represented evidence which led to the formulation of an accepted scientific concept.

Theory has to fit the evidence – so what is that evidence?

4.1.1   Origin & Evidence for the Big Bang Theory

The Big Bang Theory was originally proposed in 1927 by a Belgian Catholic priest who was also a physicist, Monsignor Georges Lemaitre[13] (University of Louvain), and was substantiated/proven by the observations of Edwin Hubble[14] in the late 1940s. The concept was further developed by the Russian-born George Gamow[15] in a paper published in 1948.

They reached their conclusions because the light from galaxies appeared to have become more red by different degrees depending on how far away they appeared to be. This is known as 'redshift'. Light comes to us in a straight line, but also as a wave. The distance between the crests of those waves is the wavelength. Light with a shorter wavelength is seen as more blue, whereas light with a longer wavelength is seen as more red.

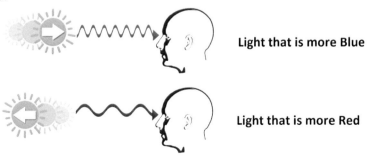

Light that is more Blue

Light that is more Red

The theory is that an object which is moving **away** from us and emitting light will 'stretch the wavelength' of the light coming in our direction making it seem more red, (a red-shift). Conversely the wavelength of light from an object coming **towards** us would be compressed or shortened making it seem more blue, (a blue-shift). This is same principle as the Doppler effect in sound where a change in tone can be heard when something loud goes past us at speed.

However by claiming that things appear 'more red' there is an implication that we know what the original light should have looked like. What allowed Hubble, (and us), to determine the original state of the light being emitted? The answer comes from spectral analysis.

When chemicals burn they emit light with strength in particular wavelengths or colours. Equally, when light passes through gasses made up of particular chemical elements there will be various points in the spectrum where particular wavelengths of light are either absorbed or additional light is emitted.

The wavelengths where light is absorbed will appear as black lines on the spectrum and the pattern for each chemical element is like a fingerprint. Observations of light from individual galaxies reveal which chemicals are present except that the absorption patterns have all been shifted by a set amount towards the red end of the spectrum.

By way of example, a normal absorption pattern within the visible spectrum might look like this

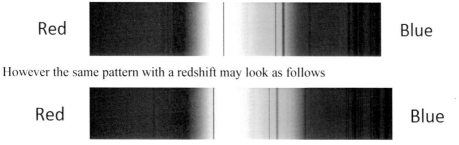

However the same pattern with a redshift may look as follows

The black lines have all shifted towards the red end of the spectrum.

Hubble was surprised to find that *all* galaxies display a redshift, (ie. they are **all** moving away from us), and he also found that without exception, the farther away that a galaxy was from us, the greater the redshift, which meant that the more distant galaxies were moving faster than closer galaxies. This reinforced ideas of an explosion.

The scientist Alexander Friedmann[16] originally speculated that such an explosion would gradually be slowed down by the force of gravity, ultimately causing the expansion of the Universe to stop and then reverse direction, before bringing everything back together again as a singularity. This 'Big Crunch' would therefore generate the conditions for a new Big Bang event; forming the basis of an eternal cycle.

This pattern of expansion and contraction, (the 'Bang-Crunch Cycle'), suggested that the creation of a Universe might be a regular event across eternity that would allow every permutation of existence to emerge at some point, as long as each Big Bang allowed nature to re-set the rules of existence. This is seen as an important factor because it allowed our configuration of Universe to be achieved without luck.

As we see stars and galaxies in all directions around the Earth, and they all display a redshift, (apart from one – the Andromeda galaxy, a close neighbour), then all must be moving away from us, but that doesn't mean that we are at the centre of the Universe. It just means that we would be somewhere in the middle of this ball.

> To make a bit more sense of this statement let me say that for all objects to display a redshift, objects behind us must either be travelling slower than us, (to generate an increasing gap between us), or travelling in the opposite direction. Conversely, all objects ahead of us must be travelling faster than us, (increasing the gap and maintaining the redshift effect).

If we imagine that material in the Universe is expanding outwards in all directions from a central point, each bit travelling in a straight line, it suggests that it might be possible to find the centre of the Universe by tracing back the path of a small but reasonable sample of galaxies, (ie. reversing their current trajectories).

I heard rumours that some scientists had attempted to do just that, in order to find the centre of the Universe and the position where the Big Bang occurred, but unfortunately I can't find any evidence for such a study and there seem to be contrasting views about whether it is even possible to do this. The core problem is that we don't really know which direction any particular galaxy is travelling in, because the distances are so great that they barely seem to move in space, even over the course of a human lifetime. The redshift can't tell us very much either.

> To understand the difficulty let's imagine that we are on a spaceship next to 3 other spaceships at the same position. Let's then suppose that we start to move at a very fast speed in one direction, and the other 3 spaceships also move at exactly the same speed but taking different courses.

> The first moves on a parallel path to us in the same direction; the second takes a course that's at an angle of $45°$ to our own; while the third travels in completely the opposite direction.

> The redshift indicates the speed we are moving *apart* so the first ship will show no redshift at all; the second would show a slight redshift because we will see it gradually moving away from us due to the angle of its course compared to ours; while the third would show the greatest redshift as it will be moving <u>away</u> from us at *twice* the speed we are all travelling because we're going in opposite directions.

> The rate at which we are moving apart is a combination of our actual speeds together with angles we are travelling at, however the galaxies we're measuring are so far away that the only thing we can be sure about is the redshift - which can't tell us either the speed or angle with any precision. The only way to gauge speed and direction for distant objects like galaxies is to observe them for a long period of time, probably longer than several lifetimes, and also to have a reference point from which to judge their movement... and sadly there are no reliable reference points.

> Neither can we be sure where the *middle* of the universe might be because we can't see its edges, and some say that it may not have an edge at all.

In short, there's a reasonable amount of evidence to support the idea of an explosive event 13.77 billion years ago, and that we are somewhere in the middle of that exploding material, but we're not sure where.

If this theory is correct, (and there are other suggestions), it would imply that the Universe is a **finite** size, (ie. doesn't go on forever), as no explosion can be infinite.

4.1.2   Dilemma Posed by the Size of our Universe

Having read this far, you may not be surprised when I say that concepts about creation and the Big Bang are severely challenged by the sheer scale of the Universe.

The part of the Universe that we can see is enormous in its own right, but we don't know where it ends. Some theories say that the Actual Universe is smaller than the distances we estimate, however most scientific theories take the view that the Actual Universe is bigger than the part we can see – the Visible or Observable Universe.

The size of the Actual Universe is therefore unknown and there's a big debate about whether it's a finite size (having a limit/boundary), or whether it is of infinite size, (without a limit/boundary). As Einstein said

> Two things are infinite: the universe and human stupidity; and I'm not sure about the universe.

As a Big Bang explosion would mean that our part of the Universe is finite then it is either expanding into a bigger and possibly infinite space, or it represents the entirety of everything and that existence stops at is edges, (ie. there is no space beyond it).

Yet we have seen the WMAP survey[1] conclude that there is probably no curvature of space which implies that it may be infinite, so we have to wonder how something infinite could exist? If it is eternal then that's just the way things are; but if an infinite Universe had a beginning that is much more challenging.

It is a moot point whether totally empty space actually represents something, (other than potential), but it seems equally possible that infinite space could hold other things than our universe. After all, if physical existence can happen here, there seems to be nothing to stop it from occurring elsewhere too. Indeed it may be that pure creation could occur on an infinite scale either as a series of Big Bang events or as one huge burst of creation – either naturally or though the actions of a Creator God.

When looking at the stars, distances are generally described in 'Light Years' (the distance which can be travelled by light over the course of 1 year). As light travels at approx. 186,000 miles per second, you can imagine that a light year is a very large distance indeed, (approx. 6 trillion miles (ie. 6 million million miles or 10 million million kilometres)). This is also deemed the maximum speed that anything can travel.

The most prominent scientifically estimated size of the 'Observable Universe', (the region which might conceivably be seen/detected by people on Earth), is at least 93 thousand million light years (93 billion light years)[16a], yet the age of the Universe is estimated to be 13.77 billion years. That gives us a problem.

Beams of light will travel in opposite directions from an open source, so if the Universe originated at a point in space, basic maths tells you that the *largest* it should be by now is

13.77 billion light years x 2 = **27.54 billion light years**   (not 93 billion).

At headline level, if the estimated size of the 'Observable Universe' is correct the most distant galaxies either had to break the rigid Laws of Physics by a considerable degree in order to travel that far, or the idea of an exploding singularity is wrong.

Put more simply, the numbers don't hang together.

To bridge the gap while preserving the idea of an exploding singularity, current mainstream theory suggests that in the first moments of the Big Bang explosion there was a period of enormously rapid expansion at speeds many times the speed of light. This 'Inflation' is deemed to have suddenly ended when all material in the Universe slowed down to just expand at the much more modest speeds suggested by the current redshifts, (even though we are not aware of anything that could have 'put the brakes on' in space).

> This 'slowing down' is quite an important point because if the theory of Inflation is correct we are missing something pretty significant out there in space. Could it again point to other types of stuff like Dark Matter which have a gravitational/slowing effect, or does it point to some intrinsic properties related to the mechanism powering Inflation – such as a Framework for Existence, (a concept we will explore in Chapter 8).

Some scientists believe that the 'mechanism' of Cosmic Inflation helps to explain why the background radiation from the Big Bang has cooled to the level that we monitor it today, within the time available, (13.77 billion years). However this is only valid if the notion of an exploding singularity was right in the first place, and if estimates about the original heat of the explosion are correct.

In terms of breaking the Laws of Physics, the theory of Inflation could undermine ideas based on a maximum speed, so scientists have argued that the speed of light is only a constraint *within* the Universe; this rule doesn't apply to its expansion. Mmmm.

Referring to the theory of Inflation in his book 'The Grand Design', Stephen Hawking described the effect as being equivalent to something the size of a centimetre suddenly becoming ten million times the size of the Milky Way, (our own huge galaxy), in a tiny fraction of a second. He then admits that... "for our theoretical models of the Universe to work, the initial state of the Universe had to be set up in a very special and highly improbable way. Thus traditional inflation theory resolves one set of issues but creates another – the need for a very special initial state."

One thing is clear from these debates - scientists themselves are not entirely persuaded by the idea of inflation. It either has to be refined with more evidence, (including a rationale for how the Laws of Physics could be broken), or we have to look at other possibilities. We will see later, (in Chapters 5 & 10), that there is one concept which may provide a way for the Laws of Physics to be broken momentarily, but it depends on hidden dimensions of existence. Equally we will see that there are other explanations which may not require the mechanism of Inflation at all.

> As a simple idea, the Laws of Physics wouldn't need to be broken if the Universe **didn't** start out as a tiny singularity but was already widely spread-out before the event. In this way the distances to be covered would be well within the speed of light, but the Big Bang would then be a very different type of event / mechanism. We will see this in section 4.2 below.

Given the many doubts that now exist about traditional Big Bang theory it's tempting to ask whether there could be a much more simple explanation?

Could the estimated size of the Universe be wrong?

That **does remain a possibility** because in recent decades we **have** seen at least one dramatic change in these estimates based on new factors coming to light... if you pardon the pun.

I will show you some factors which may be vulnerable in this respect after setting the context a little more.

Measuring the size of the universe is not an exact science. We can't take a tape measure to a nearby galaxy and then take a reading. Neither can we transmit a beam of light or a radio signal beyond our solar system and expect to see & measure its reflection. The distances are simply too great... as we can see from the length of time it takes for signals to reach us from our own probes near Saturn or Pluto.

As there are no objects in the distant universe which we can use as a fixed reference point, we either have to estimate the distances of objects using trigonometry, or generate estimates from the way that light changes or radiation decays.

In addition we don't know how fast our solar system is travelling through space, or indeed whether 'Inflation' is still happening to some degree, because the redshift evidence only gives us a yardstick compared to our own movement, and that could reflect different factors.

> For instance, a particular level of redshift could be the result of an object travelling at high speed at an angle of $90°$ to us; or an object travelling at much slower speeds at $180°$ to us. Indeed, the object may not be moving at all – **we** may be the ones travelling quickly... and reality probably reflects all 3 factors in different strengths/combinations.

So what gives us our sense of scale?

Firstly, there is nothing to indicate an outer limit because it is still possible that the Big Bang was an event within a larger Universe, so we cannot say that 27.54 billion light years is the maximum size of Universe.

Neither can we assume that 27.54 billion light years has actually been achieved from a tiny start point because it would mean that the physical material of the Universe, (including us), was actually moving outwards at the maximum speed possible – the speed of light. Within the rules of existence, (the Laws of Physics), it is not only more likely that we are travelling more slowly than the maximum speed, (which would make the diameter of the universe smaller); but if we were truly moving at the speed of light then most of the light from the exploding universe could never catch up with us.

A similar point arises from the way we define the Observable Universe. If we imagine that an infinitely large Universe was suddenly put in place as we see it 13.77 billion years ago and none of the galaxies within it were moving, then 27.54 billion light years is the maximum that we could see of it when we look in opposite directions into the night sky. If there was anything more distant then we could never see it because the light could never reach us in the time available. From this it is clear that the amount of material we are able to see is directly linked to our estimates of its age.

So if this type of reasoning can't determine whether the Universe is larger or smaller than 27.54 billion light years we have to use other criteria, and in crude terms there are 4 main factors that scientists have used :-

i. the density/heat of background radiation, and estimates of the way in which it might decay/cool, to give us an estimate of the **age** of the Universe;

ii. applying trigonometry to the position that we see objects in the sky, via a principle called 'Parallax';

iii. the luminosity, (ie. brightness of the light we receive from distant objects), on the basis that the brighter something is, the closer it must be.

iv. the degree of redshift that we observe

These and other fine tuning techniques are used in various combinations to form what is known as the 'Cosmic Distance Ladder' (or Extragalactic Distance Scale). They are all dependent on the accuracy with which we calibrate our equipment. If any of them turns out to be wrong then they would greatly affect all of the estimates, and this **has** happened in the not too distant past.

Luminosity is the measurement of how much light we get from an object in a given period, (how bright it appears to be). The principle is that the more distant an object, the more its light will have spread out, becoming fainter. **However for this principle to work we need to know how bright an object was in the first place, close-to.**

The chosen method for calibrating our scale of luminosity is to look for 'Type 1a Supernovae'. A supernova is an incredibly powerful burst of light when a dying star explodes. It will last for about 3 weeks, and will be far brighter than the entire rest of the galaxy in which it sits. Although stars have a lifespan of several billion years, there are so many stars in the universe that such events occur on quite a regular basis somewhere in the Universe – we just don't know where, but in the 3 weeks that they're visible we should be able to detect them.

In terms of calibrating luminosity, three of the biggest assumptions made by astronomy are that:

- all 'Type 1a Supernovae' are believed to have the same brightness, **and**
- such brightness has been determined for objects a certain distance away.
- Cepheid Stars tend to have a pulse that varies directly with their luminosity.

You don't really need to understand the technicalities of these points; only that these and other assumptions are not necessarily guarantees. In the 1950s the German astronomer Walter Baade (working in the USA) concluded that there were actually two types of Cepheid star, and that the calibration of luminosity had been previously undertaken only on a Type 1 basis. As a result, distances for many stars, (and indeed the estimated size of our own galaxy, The Milky Way), were instantly doubled.... not an insignificant change!

Having said that, science uses the different techniques it has in its armoury to cross-check its findings as best it can, and in this respect it was reassuring that NASA's WMAP programme did reinforce a lot of the current estimates when it scanned the universe for background radiation in 2001. This essentially used estimated decay rates in the Cosmic Background Radiation and it confirmed estimates of the age of the Universe at 13.77 billion years.

The other major technique that I mentioned above is Parallax, which works on the principle that if you know the angles of a triangle and the length of one side, then you can work out the lengths of the other sides... the distance to far-off objects.

In astronomy these are very stretched-out triangles where the distance to another galaxy or star will form the long side, while the short edge depends on whether we imagine **ourselves** to be at the most pointed end of that triangle or whether that sharpest point is situated at the **object** we are observing. As examples :-

- If the short edge of the triangle was far away it could represent the distance between two objects that we see, or the distance that an object appears to move if we make observations on different days.

- Alternatively the short edge may represent the movement that **we** have undertaken between different observations, given that we know the size of our orbit around the sun and how far we will have travelled around it.

Astronomers have very sensitive equipment to measure angles, and from these types of calculation we can form an impression of distance, but for accuracy we need to calibrate distances using some of the other techniques as a yardstick.

Science has put a lot of care into calibrating its findings, so while we may keep a wary eye on developments, the current estimates are the best description we have available from the observations made. However if we do trust the resulting numbers and believe that the size of the Observable Universe is far bigger than 27.54 billion light years, then we are faced with some uncomfortable realities.

Using the Cosmic Distance Ladder, the most distant objects/galaxies we can see are thought to be approx. 13.2 billion light years away. In other words, the positions where we see them today would have been achieved 'just' **half a billion years after the Big Bang event**.... and they are widely spread out. In crude terms for instance, if the Universe post inflation cannot expand faster than light then its maximum further spread would be 27.53 billion light years. Knocking this off the 93 billion figure we arrive at a size that is in excess of 65 billion light years in those early days.

This is one reason why many scientists argue in favour of the theory of Cosmic Inflation. Another reason which is often cited is that it helps to explain the density/heat of the Cosmic Microwave Background Radiation (CMBR) - but I feel this is a circular argument because the two factors are directly linked via the age of the Universe.

Yet if you are not persuaded by Cosmic Inflation you have to explain the size & redshift evidence in other ways. Some people will 'embrace' the larger estimates as evidence of an infinite and eternal Universe, while others will seek to reduce the size in order to bring it into line with the age of the Universe, for instance:-

> If light has taken 13.2 billion years to reach us, it doesn't have to mean that the object/galaxy was **originally** 13.2 billion light years away. The argument runs like this:
>
> If we imagine that our own galaxy (the Milky Way) is travelling on the same course as the other galaxy but we are ahead of it and also moving faster, the original distances between them could have been a lot smaller. If we are travelling at 85% of the speed of light and the other galaxy is at 10% of the speed of light, when the original burst of light was emitted we would be moving **apart** at $^3/_4$ the speed of light, but equally light would only be catching us up at $^1/_4$ the speed of light. If we were originally 3.3 billion light years apart that light would take 13.2 billion light years to reach us, and we would only just be 13.2 billion light years apart now.

I am not entirely persuaded by this. Although I can see the basic principle, this notion is based on many assumptions, and we can't be on the same trajectory as **all** of the most distant galaxies which are widely spread out. I am merely pointing out the range of thinking. Equally, people who believe in Inflation must find a credible way to explain how it could be possible. Until then many others will remain sceptical about it.

4.1.3 Another Problem

If the size of the Universe and its lack of curvature were causing some discomfort in the scientific community, you will now appreciate the shock which a recent discovery has generated, as it further undermines the Bang-Crunch theory – and in a serious way. It was announced in the usual understated manner:

Redshifts in type 1a supernovae are increasing[17].

Layman's translation : We have seen, that Type 1 a Supernovae have been used as a way to calibrate our estimates of distance. These measurements show that the observed redshift of these distant events is becoming more pronounced which means that the expansion of the universe is underlined[17], **not** slowing down as expected.

Scientists also believe this acceleration began to happen 5 Billion years ago – not from the start of the Big Bang[41].

Because the Universe is not slowing down, it cannot collapse in on itself, so **the theory of a Bang-Crunch cycle can no longer continue to apply**.

In relation to the Bang-Crunch theory this evidence tells us that the Big Bang was either a one-off event, or that our Big Bang marked the *end* of a previously eternal 'Bang-Crunch cycle'. In either of these circumstances **an external or spontaneous factor would be necessary to make the change**.

Yet there is supposedly nothing else beyond the Universe, (the totality of everything), and Matter/Energy isn't capable of being spontaneous.

This has profound implications because the main materialist explanation for the origin of existence has been fundamentally undermined in several ways.

Firstly, the whole purpose in suggesting an eternal cycle of Big Bangs was to allow nature to explore every permutation of reality in order to avoid luck in bringing about our version of existence. If the sudden acceleration was an inevitable event due to strict cause & effect, then the Bang-Crunch sequence might occur a few times but it could **never** be eternal and it would therefore fail in its prime purpose.

On the other hand if the Bang-Crunch cycle was potentially eternal but had been ended by an external factor, it now needs something to not only break the cycle but also to do it **only after** the right conditions for our Universe were achieved if we are still to avoid luck. That would almost certainly require design and co-ordination.

To complete this avenue of thinking, if we say that there is nothing beyond the Universe, (no external factor), then the Bang-Crunch explanation **can only remain viable with spontaneity/Thought**.

You can see why Materialists & Determinists are keen to find new explanations of origin which can preserve their views of existence. This is where we begin our exploration of other ideas.

However there are some practicalities too. If we liken the acceleration to making our cars go faster, we know that more petrol is used when we press the pedal. The engine converts the petrol into energy which raises the speed. When we stop accelerating the car continues at the new speed until something else slows it down, (eg. air resistance).

In the same way, we need to find a new source of energy to power the acceleration in the expansion of the Universe. Put another way, there is no air resistance to slow things down in space, but there is still a need for energy to raise the speed.

The amount of energy required to achieve this effect is truly vast. NASA calculates this 'Dark Energy' to be approximately 15 times the amount of matter represented by all the atoms, (galaxies, gas clouds etc), in the visible universe. That energy would have to be spread evenly everywhere in the Universe to generate a uniform/balanced effect, yet we do not detect it.

If the Big Bang initially released raw energy, (as is commonly believed), then we may assume that any energy which was not converted into physical matter would be used to drive the expansion, (ie. the original flat speed that would presumably have applied for the first 8.77 billion years after the event).

To cause an acceleration a new source of energy either has to convert some of that physical material back into energy, or there must be an external source.

Given the timing of that acceleration, (beginning just before the formation of our solar system), it's inevitable that people will wonder if the two events are linked and what the external source might be? There are only a small number of possibilities.

4.2   Are There Different Interpretations?

Science **has** detected a number of things which point to an event 13.77 billion years ago. Something seems to have happened at that time, so if we are not persuaded that it occurred as a result of an exploding singularity we must consider other options. In this section I will set the strategic context by exploring a couple of possibilities.

We have seen 4 key characteristics in the spread and pattern of movement of galaxies across the cosmos, which have to be explained:

- There is a very even distribution of material across the visible Universe.
- The light from almost all galaxies shows a redshift, which means that they are moving away from us and probably outwards from a central point.
- The more distant the galaxy, (measured by the faintness of their light), the more pronounced their redshift becomes. So the closer to the 'edge of the Universe' that they are, the faster they seem to be travelling away from us.
- Redshifts are increasing – so the expansion of the Universe is accelerating, not slowing down as expected, and this seems to have begun 5 billion years ago.

These patterns **are** very reminiscent of an explosion and in broad terms there only seem to be two types of effect which could generate all aspects of this activity :–

a) an outward push from within the Universe, (generated either by distortions in the fabric of space; one or more explosive forces; or some sort of pressure),

b) an absolutely uniform force of attraction/suction surrounding the Universe.

Scientific thinking about how the pattern of movement was achieved has been dominated by option 'a', because the general perception has been that the Universe represents the totality of everything, so whatever is causing it to expand must come from within - there can be nothing beyond it to create any influence.

To be more specific, there is no known 'suction' force which emerges from a void.

Science tells us that the suction which we perceive when air rushes in to fill a vacuum *isn't* a force of attraction generated by the void, but the force of the air pressure around the void. Similar types of effect always come from places where there is more existence, and in the context of the Big Bang that would be the Universe - not any void around it.

So if we believe that the material of the Universe is being pulled outwards, it is more likely to represent a gravitational effect from something which lies beyond the Observable Universe, (such as scattered chaotic material, or other 'Universes').

If this was correct, then the galaxies on the outer edges of the Observable Universe would be the closest to this other material and therefore the most influenced by its gravitational attraction, while galaxies closer to the centre of our Universe would feel a weaker influence. So while this does seem strange at first hearing, it **could** potentially generate the same 4 symptoms listed above, although it would have to be a very uniform and balanced force – meaning that the unknown material was very evenly distributed around us.

As an explanation for the Big Bang there are several downsides to this thinking. The gravitational effects we perceive from distant galaxies, (that we can see beyond our own galaxy), appear negligible, so it is hard to imagine that material even farther out, beyond our view, could exert the necessary level of influence to pull our galaxies outwards mimicking a single Big Bang explosion, when we can't detect that influence.

It would have to be a very gentle effect in order to 'slip below our radar'; and to achieve the size of our Observable Universe within the Laws of Physics, the wider infinite Universe, (**ex**ternal gravitational effect), may have had to work on physical material that was already widely spread-out 13.77 billion years ago, which was more concentrated than the background universe. This would mean that our space was a distinct bubble of activity... but why? A new burst of creation, or a very large object drawing things towards it before it exploded?

In one sense this general effect is feasible because we don't even detect the gravity of our own galaxy, which should be a lot more noticeable. We could liken the influences to water moving down a slope - say an aqueduct with just the slightest of inclines... however this would imply a steady build-up of speed from the earliest moments after the Big Bang, while science tells us that this effect only began 8.77 billion years later.

On this basis such an effect may be more likely to explain 'Dark Energy', (including why we haven't yet detected that either), although we would also have to explain why the effects suddenly kicked-in 5 billion years ago. Could the outlying material suddenly have become a lot closer – allowing our Universe to suddenly feel its influence?

To me, aspects of such scenarios seem far-fetched because they don't seem viable in isolation, (such as explaining the bubble of activity). They also fail to offer any apparent prospect of an 'eternal cycle', and would therefore make them dependent on luck to achieve our configuration of reality, which isn't generally considered acceptable.

I have mentioned them because they contain interesting features which we may find useful elsewhere, and because they do offer a strategic line of enquiry. Yet it would be a lot simpler for Materialist theory to directly tackle the issue at hand, by finding another cyclical mechanism of origin that is not affected by the accelerating expansion.

That is precisely what it has done.

A prominent version of this thinking suggests that two huge membranes/**curtains of energy**[38] exist beyond our field of vision. The suggestion is that if the curtains touch they would release vast amounts of energy in an alternate form of Big Bang which would generate a new Universe. This would represent the start of a new cycle.

The physical matter which crystallised from this burst of energy would form a Universe which would push the curtains apart, preventing further Big Bang releases until the material became so thinly spread by its expansion that it could no longer act as a barrier. The curtains would then come together again and generate a new Big Bang, establishing a potentially eternal cycle that is unaffected by expansion.

This genre of thinking can be referred-to as 'Brane Cosmology', and within this, the particular model that I have described is known as the Ekpyrotic Universe[38]. This type of theory expects that all Universes will expand and dissipate to near nothingness and will never collapse as a Big Crunch.

It's a neat idea which overcomes several problems in the Bang-Crunch theory but is doesn't escape the need to explain why our expansion is accelerating. To this end people have speculated that a **partial** release of energy from these hidden structures/membranes/curtains might be a source of Dark Energy.

We will return to Brane Cosmology in later chapters, however, given the many layers of speculation within this type of thinking, those with a religious disposition might sigh at the level of criticism they have faced on far more simple suggestions !

4.3   Other Events Like This

As mentioned earlier, the evidence for the Big Bang strongly suggests that there was an explosion, as a way to explain the pattern of movement in the galaxies of the Universe which astronomers have observed. It is the most simple and obvious explanation even if there are other theoretical ways to explain the patterns of movement in the cosmos. My final points in this chapter therefore consider whether there is any evidence for other Big Bang creation events?

4.3.1   How Many Big Bangs might there have been?

The Bang-Crunch theory envisaged a perpetual series of Big Bang events as a way to avoid a true beginning to the Universe and to enable nature to explore every permutation of reality. The implication from the accelerating expansion of the Universe is that our Universe will not be able to collapse again into a singularity, but this doesn't necessarily disprove the possibility of earlier Big Bang events, (prior to 13.77 billion years ago). It could 'just' mean that we are the last of the cycle, or possibly that other mechanisms were at play, (on the lines of Brane Cosmology above).

There are some types of radiation, (such as the Cosmic Microwave Background Radiation - CMBR), which are believed to never disappear but which are supposed to only emerge under very specific conditions such as a Big Bang.

As we know, the WMAP[1] programme mapped the density/heat of the CMBR across the Universe, to determine its pattern of distribution, and people searched for any signs of gravitational waves or even earlier Big Bangs.

The Planck space telescope, (launched in 2009 by the European Space Agency)[18], was also scanning the heavens doing precisely this until 2013.

Final results were expected towards the end of 2016, so I was hoping to include them in this book, but I haven't seen them. However my understanding is that while numerous small-scale concentrations of radiation have been detected, they are spread evenly across the entirety of the Universe with nothing to suggest an overlapping event on the same scale as the Big Bang.

Assuming that this holds true then theories of origin have to accommodate the absence of evidence for such earlier events. For instance, Bang-Crunch would have to argue that all radiation would be returned to the singularity, clearing away all evidence of earlier events, while the curtains of energy approach would say that all residual radiation from prior events would have been pushed well beyond our sight.

What do you think?

4.3.2   Evidence for Spontaneous Physical Creation?

In broad terms, science says there is no evidence of physical matter/energy just popping into existence. (If there were, then some aspects of physical existence would have to break the principle of cause and effect).

Yet the notion of Spontaneous Pure Creation for physical Matter/Energy remains popular, which is why Big Bang theory forces science as a whole to be non-committal about what that event may have been – even if individual scientists are biased one way or the other. However to say that there is no evidence of spontaneous physical creation may not be entirely correct. It may be a question of interpretation.

While it is true that we have no evidence for overlapping Big Bang events, astronomers do observe really bright explosive events (supernovae) on a regular basis. These are always *interpreted* as the death of stars, spreading their residual material in all directions, because it is assumed that spontaneous pure creation is impossible.

Yet it may be that some or all these are in fact new bursts of creation, and that the spread of hot material from these explosions is brand new matter. To my knowledge we have never witnessed the triggering of a supernova – we have only spotted its bright light after the event. Neither do we have an ability to measure the totality of existence to see if it is changing, or even to see what preceded each supernova to determine what changed.

In large part, it is the presumption of causality that provides the dominant viewpoint.

Re-interpreting the nature of supernovae events illustrates how ideas about a single point of origin may be wrong and that there may have been multiple start points or even a constant trickle of new existence, if spontaneous creation was indeed possible.

It *is* possible to explain the even spread, and pattern of movement that we see in the cosmos from multiple start-points as we will see in later chapters.

We can therefore see how assumptions can colour perceptions.

In fairness, there is **no** evidence that the physical material we normally experience is able to do anything spontaneous or random, so the assumption that a supernova represents the death of a star is not an unreasonable assumption for such a fundamental point. Yet we should remember that it **is** a theory, and that supernovae are extreme events - not the cosy norm in which scientists test the abilities of Matter /Energy.

Ideas concerning spontaneous pure creation are continued in Chapter 11.

5   Scientific Views of Existence

> I always avoid prophesying beforehand because it is much better to
> prophesy after the event has already taken place.
>
> (Winston Churchill)

It's the way of the world that when people do not have an answer, they speculate or prophesy, and more often than not they get it wrong. This applies to science as much as it does to religion, however there's no harm in forming these ideas when they're declared to be notion rather than fact. Unfortunately those boundaries are often blurred.

I suspect that this is driven by the pressure to find explanations, which is as strong today as it ever was; it's just that the unknown factors now lie much deeper in the recesses of existence. They are therefore more removed from the surface levels which we inhabit, making them harder to identify and explain.

Through this, the nature of 'challenging the unthinkable' has also changed its face.

The pillars of belief about our existence are still founded on things that we cannot see, and can only speculate about. People put proposals forward as unshakeable beliefs only to find that new evidence proves them wrong and the shocked community has to think again. (A recent example was that the expansion of the Universe is not slowing down as previously believed - it's *accelerating*).

To understand the significance of new findings we have to delve into the heart of these subject areas, but rest assured, **I will try to do so in the least technical way so that everyone can follow the logic**.

The purpose of this chapter is to show the range of scientific thinking about the underlying nature of our reality, and at high level, to show where the strengths and weaknesses of these ideas lie.

In overview, many of these notions come from manipulating equations because we cannot directly see what is going on in the core of existence, not even with the most powerful microscopes available to mankind. So let's begin by setting a context.

Approximately 130 years ago the scientific community began to realise that it had to split its explanations of physical reality into two parts:

- What happens at the level of **atoms, molecules and larger objects,** (the level of the reality we occupy, which is the *traditional science* of Newton & Einstein).

- What happens at the deepest levels of existence, *within* atoms - the actions of much tinier '**sub-atomic**' **component particles** (the new science of *Quantum Mechanics* advocated by people such as Heisenberg, Bohr, Schrödinger, Fermi, Planck and many others).

Traditional science can't explain how sub atomic particles are behaving. Some of the genuine findings from Quantum Mechanics are truly bizarre, and seem impossible. In other experiments exactly the same test can produce a great variety of results – thereby appearing to break the principle of single cause - single effect.

They appear to have their own separate Laws of Physics, and while both levels of existence must work together to form a single reality, the different narratives have remained largely separate even to this day.

The one crumb of comfort is that sub-atomic particles still seem to follow loose **rules**, because the variety of outcomes still tend to fall within a range, and that implies a degree of structure plus the hope that the principle of Causality can be salvaged.

As important examples of the 'bizarre' which have also been repeated many times, two discoveries stand-out for me, and these have driven a lot of debate because they are not factors that could be explained as false results from poor or insensitive equipment. They are more direct findings which will be explored in chapter 6, but in summary:-

1. The simple act of closely ***observing*** the behaviour of sub-atomic particles causes them to act differently and even to change their state of existence.

    These remarkable findings from the 'Dual Slit' and 'EPR' experiments, (in their various forms), are discussed in the next two chapters, and they began to unravel the traditional scientific view of physical existence.

2. Sub-atomic particles which are 'paired' appear to retain a link even when they are moving apart at twice the speed of light and are separated by large distances – as evidenced by them physically influencing each other instantly – a level of communication that is at least ***10,000 times faster than the speed of light***.[19]

Both of these factors should be impossible according to traditional thinking, yet here they are. The experiments have been verified, so the underlying facts are not in doubt, but we lack explanation and the obvious implications drive a passionate debate over the nature of existence and the hidden influences which might control particles.

To explain the first finding a little more, observation is a process that 'receives' – it doesn't transmit anything physical. So when we find that sub-atomic particles such as photons and electrons change their behaviour when they are simply observed closely, we have to consider whether they are responding to something intangible. This is why some people feel these experiments are evidence that the analytical thoughts of the observer are generating physical effects beyond the brain, (ie. mind controlling matter). As you can imagine, this is highly controversial.

The 'Wheeler Delayed Choice' and 'Quantum Eraser' experiments (Ch 6) have sought to probe these findings more deeply and to trick nature into revealing more about what is happening. Interestingly, and contrary to traditional scientific expectations, they did **not** end up removing the suggestion that Thought may have a direct influence on matter/energy. You will have to make up your own mind about these experiments later.

While there ***are*** theories that try to explain the growing list of such findings **without** Thought, they don't explain all of the significant observations which have been recorded. On the other hand, when people do use Thought to fill the gaps in what is happening, and portray this stuff as 'something' in its own right, (ie. having capabilities beyond the physical brain or even beyond a chain of physical events), the arguments are immediately undermined because we can't point to Thought and say 'here it is'. We **can't** look at electrical signals in the brain and say 'electricity thinks', so there's no way to establish a direct link between Thought and the experimental results.

As a result, while there are partial explanations on both sides of debate, (Matter vs Thought), neither perspective can claim to provide a fully robust solution.

In relation to the second numbered finding above, it has been a long-standing principle in science that nothing should be able to travel faster than the speed of light, yet the ability of particles to instantly *influence* each other when they are moving apart at twice the speed of light has been demonstrated and measured at the ends of a cable 10km long. It suggests hidden mechanisms of control, (see 7.1.2, 9.1.5, and 10.2.1).

Having set the context, the evidence I will present in Chapters 5-7 deals with what <u>does</u> exist at the core of our physical reality, while Chapters 8 & 9 cover the evidence for certain additional features that *may* exist.

For the purposes of our debate, these findings give us pointers to what may have originally emerged from the origin of the Universe, but before delving into the details I want to take a short diversion to provide additional context for some current theories.

5.1   New Types of Equation   & the Ball of Possibility / Time

Relax! Be cool. I'm not going to be throwing mathematical formulae at you!

I can almost feel the angst as many of you read the title of this section - not least about the looming dullness! But don't worry, these few pages will hopefully be interesting as well as requiring some mild 'mental gymnastics' concerning ideas about Time.

Within its four and a half pages I will provide some justification for earlier comments about the limited usefulness of equations and why we should be cautious about what they may tell us. Along the way I will also introduce you to some concepts about the possible structure of the universe and Time which have emerged from the manipulation of formulae. These ideas will underpin later debates.

As we know, modern science carefully observes reality and then tries to find ways to describe it. From this scientists have found principles that can be regarded as 'Laws of Physics' which describe how nature operates on a rigorously consistent basis. These rules have been translated into mathematical equations/models however once they are encoded they can be interpreted in new ways, and through this process they can sometimes be separated from the original notion by incorporating new assumptions.

For example let's look at Stephen Hawking's[20] description of the expanding Universe.

> His concept of the expanding Universe is like an inflating ball, in which he argues that the space between galaxies is increasing as material moves outwards, yet the galaxies themselves are not expanding.
>
> As a mathematical way to describe an explosion this can be a useful clarification and at headline level it would seem to be true but the statement covers up some contentious points that have implications for our thinking.
>
> Because nothing is believed to exist beyond the Universe, the idea has grown that 'space' has to be created before physical matter can occupy it. Space is seen as something that exists, (either having a 'fabric' or being like a Framework for Existence). This idea has largely been proposed to avoid suggestions that space, as a potential place of existence, could be infinite – which no explosion or other physical mechanism could create or fill.
>
> Once in the public domain such mathematics has been used to imply that this **finite** framework is being stretched... yet space *could* be just an empty void that is infinite.

How we **use** such equations is therefore important because although maths is very helpful it can incorporate assumptions that are not universally accepted. On matters of prime origin and the unknown, we have to be particularly careful when using them.

At a strategic level, 'equations' incorporate 2 principles :

a) 'cause and effect' which, according to science, is a basic feature of physical Matter/Energy, and also

b) that nothing is ever created or destroyed, only transformed.

Equations can only work if one side of the formula exactly equals the other side. In other words, if the behaviour of **Matter/Energy** always conforms the relevant equation, the inputs to any process must always exactly balance the outputs.

In addition, if correct, Matter/Energy should <u>only</u> be able to work *within* the principles/rules encoded in the equations and **will not be able to rewrite them**.

This causes a problem when we consider that the Big Bang either seems to represent a major re-write of the natural laws, or a moment of pure creation, (both of which seem to require random or spontaneous acts which are not possible for matter/energy, and which could also break the principles within any equation).

Put another way, if something could suddenly pop into existence, (pure creation), then mathematically it would be the equivalent of 'zero suddenly becoming 1' for no apparent reason. A true creation event would therefore mean that **relevant equations couldn't balance or exist**.... and we can't have that, can we... unless it's true.

As pure creation is one of only two options for prime origin, and the concept remains very popular especially in the light of the Big Bang theory, science and atheism have searched for ways to bring it into a balanced equation, and the main approach is to suggest that in a creation process nothingness will generate balanced quantities of 'positive' and 'negative' matter. As Einstein said, all-be-it in a different context

Creativity is knowing how to hide your sources.

Traditional theories and equations were all formulated on a similar/compatible basis, allowing scientists to combine those from Newton with others from Einstein etc, In this way new and deeper insights could be derived about nature. Perhaps the best example of this was Einstein's equation $E = MC^2$.

Yet such an approach has clear dangers when equations are based on 'nice ideas' & speculation rather than observed reality, because if equations balance they will inevitably suggest that any idea encoded within them is possible... even if they're not possible at all in reality. This is a particular danger in equations that concern prime origin because there is no basis on which to verify some of these factors. We have to be very clear about the basis of any equation before we give it any credence.

Put another way, Materialists wouldn't accept God being factored into an equation as it would inevitably imply that He existed, without providing additional proof of this.

A step change in that danger level occurs when people try to combine traditional equations with those that have **not** been formulated on the same basis, (those from Quantum Mechanics), as some scientists have begun to do in recent years[20]. To understand this comment we need to appreciate why the approaches are so different.

The operations of objects smaller than an atom, (the 'sub-atomic' level of existence), *cannot* be described using traditional formulae because there can be a <u>range</u> of possible outcomes from a single type of event, and that breaks the principle of causality. Neither can we search for distinguishing factors because we can't see what is happening, even with the most sophisticated microscopes/equipment yet developed.

As a result a new breed of equation has emerged that can accommodate the uncertainty where we cannot observe activity precisely. This is the technique used by Quantum Mechanics, and as I've already indicated, it's based on the principle that all conceptually logical outcomes are possible but with different levels of **probability**, and that's often a very big range.

By providing a single scenario with many outcomes, (and knowing that this is true in reality), many students and an increasing number of scientists have started to say that randomness exists at the sub-atomic level of existence, however this was not the original basis on which the maths was formulated. To this day we find senior scientists trying to hold the line by saying that Quantum Mechanics still operates on the basis of causality – it's just that we don't know what the missing factors are.

When using Quantum Mechanics scientists often begin by saying that almost everything is possible from a given scenario, which is generally very unhelpful because you would never know what to expect as an outcome. The second stage of thinking is therefore to narrow-down the options by concentrating on those outcomes which have a 'reasonable' probability, (for instance by excluding those things which are deemed to have zero probability or infinite probability). The concepts that you're left with form a range of **likely** outcomes, although some odd-ball events will occasionally arise in experiments.

Quantum mechanics is not only great in theory, it has also helped us in many practical respects because it's mathematical models of the sub-atomic universe have started to allow us to predict outcomes in reality when it wasn't possible to do so beforehand. Yet its conclusions hinge on the judgements being made in whittling-down the options or assigning probabilities.

In this respect, the desire of some scientists to reduce the options can lead them to discount certain possibilities because they assume that they can't be right, however in extreme circumstances such as the origin of the Universe that could be a mistake.

Clearly, there's no harm in trying these things out because it will help us to refine theories and come closer to workable models that are more precise. The danger lies in believing that they are always accurate.

So we now come to the point where techniques appear to be stretched to their limit.

The 'Holy Grail' of physics is to achieve a single theory of existence that reflects a complete picture of what is happening. It needs to meld the theories/equations of Einstein & Newton with those of Quantum Mechanics, (which are not currently seen as directly compatible).

One attempt to unify these branches of science emerged from the realisation/assumption that in the first moments after the Big Bang activities at the sub-atomic level of existence could be applied to the whole universe, because it was supposed that only free floating energy and sub-atomic particles would exist at that time.

This led to attempts by some scientists, including some famous figures, to combine the equations of the two disciplines[20], (traditional science and quantum mechanics), which as we have seen, contain significant areas of incompatibility.

At this point science entered the realm where theory about our ultimate origin is being driven by speculative equations that were not necessarily aligned. It therefore had to proceed with caution.

When Quantum Theory is used to explore explanations for the origin of our Universe the logic tends to begin by saying that at the moment of the Big Bang all outcomes were possible, (only limited by the underlying capabilities of the stuff of existence, which we don't know – although they are assumed to be very broad).

After the trigger moment of the Big Bang the way in which the universe unfolded is generally discussed by scientists in three ways. The first group says that only one path was available due to cause and effect. The second group considers that many paths could have been taken but for some reason only one path was followed. The final group suggests that all theoretical paths were available and that they are <u>all</u> being pursued in parallel, now – a very similar notion to Quantum Mechanics.

There is scientific talk of :- additional *dimensions* to our physical universe; *parallel universes* within those alternate dimensions; *multiple universes* (side by side within our own 3 dimensions); and other weird stuff… all driven by equations which supposedly give them validity. Most seem to be promoted by devotees of Quantum Theory. Yet we should always try to relate these ideas back to physical reality, eg.

> Is there any evidence for an infinite number of parallel versions of **you**, all currently leading distinct versions of their lives having made different choices at different points?

Some of these ideas will undoubtedly have greater credibility than others, but if atheistic science dismisses the religious idea of a single 'spiritual realm' it seems bizarre for it to then support concepts where potentially millions of other realms are existing side by side.

There are various reasons why people feel that multiple environments are an attractive prospect. Some may draw parallels with their religious beliefs, (eg. Hinduism and Buddhism). On the scientific side, 'multiple' or 'parallel' universes are another way to suggest that nature has the opportunity to try-out every permutation of existence, thereby guaranteeing that any particular outcome that we seek will eventually be created. As you may recall, such mechanisms avoid the need for luck when trying to explain how something specific and complex could emerge without design.

Another concept is that a series of universes having different configurations, are supporting each other in a necessary circle of existence, (to be discussed in 9.1.3).

Having come to this point, I think it's worthwhile drawing-out a particular association between the two sets of ideas, which leads us to the notion of a **Ball of Time** or a Ball of Possibility.

Normally we consider any single path from 'a' to 'b' as a line with a start and a finish. Yet if we start at one point and follow a curved path, we may end up drawing a complete circle or hoop. So one way to argue that existence had no beginning would be if Time itself were structured in a loop.

> Events that occur in our lives would sit on the timeline, but the sequence would have no beginning or end.

Yet a hoop, being narrow, suggests that there is still only one way for events to unfold – because you will always return to any point that you choose to start at, and then replay the same events again. Equally, for this to work, Time would need to have an **actual structure** which could enforce a circular path. The underlying question is whether events sitting on the Timeline were 'potential', or real and could be visited?

If you wanted to accommodate more ways in which reality might unfold, then your concept would effectively need to broaden the thickness of the hoop to allow more possible events at any point along the timeline. If we needed an infinite number of outcomes, people have imagined spinning that hoop into a ball to show how all of these different circular paths could sit next to each other[20].

Put another way, if the ball of time held an infinite number of events within it, there could be an infinite number of paths that could be taken around it – each representing one version of the life of existence.

On this basis it can be argued that each person's consciousness is simply travelling along one path around the ball of infinite possibility. The choices that we make could move us onto a different path around the ball. Alternately, there might be several versions of our consciousness travelling around the ball at the same time, taking different routes! Yes, I agree, it is bizarre... but you can hopefully see the logic.

These are strange notions which may take a while to get used to, but as Quantum Mechanics imagines that all outcomes are possible you can perhaps see how the Ball of Time idea fits-in with quantum equations. However just because such theories can be described mathematically doesn't mean that they have any greater credibility than religious speculation. Put another way, they represent another faith.

Some atheists have tried to argue that if there was no beginning or end to Time there would be no need for a God to kick it off, however you could just as easily argue that if Time did indeed have a structure, it may have to be put in place by something... God? To round off this mild diversion, we might simply say that Voltaire took the arguments to their limits when he said

> God is a circle whose centre is everywhere and circumference nowhere.

Hopefully these points show the generic concerns over equations driving perceptions of reality. It's all too easy for people to push the argument too far, losing sight of the fact that equations need to match the evidence. This is especially true of Quantum Mechanics which exists to provide some structure around circumstances where we can't actually observe what's going-on.

5.2 The Fundamental Elements of Existence

By searching for the most fundamental components of existence we're not only seeking a better understanding of how to utilise our environment, but for the purposes of our discussion are also trying to establish which factors in existence *may have had no beginning* or *emerged spontaneously out of nothing... if any.*

I therefore want to focus on the core elements of reality, and it's therefore natural to look for the things which actually give physical **substance** to existence to demonstrate that it is not just in the mind.

Einstein's equation $E = MC^2$ shows us that mass and energy are interchangeable and most scientists will interpret this to mean that the tiny 'solid nucleus' at the heart of every atom can be converted entirely to energy... a lot of it! This of course, also implies that something had to compress all of that energy into 'mass' at some earlier point as well.

A second finding from science also begins to focus our attention. Science says that atoms are actually made from a few miniscule particles in a lot of near-empty space, all bounded by force fields... whatever they might be.

Invisible forces provide structure, as well as generating chemical reactions that : bind atoms into molecules; cause the sun to shine; objects to heat-up; and other physical influences between different objects. In Chapter 2 we saw that science has identified just 4 forces:-

- 2 forces which exist at the level of *atoms and larger objects* - Gravity & Electromagnetism
- 2 which exist *within atoms*, at the sub-atomic level - the 'Strong' & 'Weak' nuclear forces.

As we will see, these factors help science to argue that physical stuff is real and perhaps it represents the only underlying stuff in existence, as believed by Materialists. On the other hand, Dualism suggests that both Thought and physical matter co-exist as different types of underlying stuff – one providing stability through cause and effect, the other providing a limited mechanism for change, not chaos.

To summarise earlier chapters, in **all** philosophies Thought provides a different type of structure through rules, concepts, or even designs – it's just that within Materialism, Thought is seen as a physical mechanism based in the inevitability of cause & effect, while Idealism and Dualism claim that it can be spontaneous & random.

It's also interesting to note that both Thought and physical Matter have static and dynamic forms. Thought is the dynamic form of 'Concept', while Matter is the static form of Energy. Such a notion suggests that movement may be caused by something else. Typically science attributes this to the 4 forces but there is another possibility – particularly when we consider the notion of a sequence – a 'before' and an 'after'.

Movement and sequence would seem to be impossible without Time, which is the rate of change, so it is equally possible to view Time as a separate type of **force**, which is a visible factor of reality.

The difficulty with Time is that we're not sure if it exists as something distinct in its own right, or whether it's just a symptom or effect of the other basic elements of existence. If it is not a force then people have speculated that Time may represent a different *dimension* of existence, however we normally consider 'dimensions' to be another aspect of structured matter, which would make Time more like a symptom.

On balance, against the range of possibilities, Time will generally be seen as an *effect* within the other generic elements of existence – but the other concepts retain validity.

So we have a basic candidate list for the underlying stuff of reality:-

- Energy, (from which mass is constructed)
- Force, (which might include Time), and
- Thought (as a different type of underlying stuff)

Because we don't know what they are, notions about Energy and Force can easily become blurred and indeed they **can** be seen as different aspects of the same thing. Yet it's important to understand the theoretical distinctions between them in case they do represent different things.

Forces pull or push, but the strength of those actions, the length of time they are applied, and the distances over which they are effective, all relate to the amount of Energy available. Put more scientifically, Energy is seen as the ability to do work, while Forces are effectively the mechanisms by which it is applied.

The next steps in our thinking become a bit more technical, but it's necessary to understand them in order to grasp some of the later ideas. I'll keep it brief and relatively simple.

How might energy and substance (ie. mass) be interchangeable? We know that if a sub-atomic particle is converted to Energy a vast amount of energy will be released and so such particles could be thought of as being highly compressed forms of energy.

At the same time, we know that some sub-atomic particles exist *which have no physical mass*, so it can be argued that *forces* are the things that capture energy in order to structure it as matter. Put another way, forces can be thought of as forming empty shells in which to contain energy.

In order to form a container, Force would have to combine power/energy with a mechanism to give it structure. Put another way, a line of force may be energy that is applied in one direction.

This evidence and logic has led some scientists to believe that the essence of existence may boil down to just one factor - Force.

However this gives us a problem. If Forces are portrayed as the natural cause of movement, how could they become static to form a structure/shell?

At this point we should again remember that we don't know what a Force actually is, yet from what we do know about the properties of Force, there are two conceptual ways in which it might become static and possibly structured: where

1. Two opposing forces meet & counteract each other – possibly holding themselves in place.

2. A single force is bent around and applied to itself, holding itself in place and/or causing it to form a static structure.

To illustrate the second point we can use the following example:

> You know that your arms are flexible and they can apply force in many directions. Equally, they may be stretched-out or be bent around to touch each other. So let's imagine that they are lines of Force.
>
> In reality, when your arms do form a circle, they become stiffer as you press your hands together with greater strength. So the flexible line of your arms can become a rigid ring, and this is how we can imagine a moving Force becoming a solid. In the same manner it's suggested that a 'sheet of force' can be bent around to form a cylinder or a hollow sphere with a hard exterior shell.
>
> > (How we get to a defined 'sheet of force' is unexplained, other than hints that lines of force may be laid side by side. We would then have to ask how can a force be a line with two ends... etc?)

Yet if this notion is correct we may begin to see how forces at the tiniest levels of existence may be shaped into containers that could hold energy. 'String Theory'[21] and its variant 'M-Theory'[22], (ie. current mainstream thinking within science), support the notion of shaped forces at the base level of existence.

> The 'Standard Model of Particle Physics' sounds appropriately grand, but is essentially a mathematical description of how the known particles in existence function and how they combine to produce the higher levels of physical matter.
>
> It's often described as 'inelegant' because of its complexity, yet it has correctly predicted the existence of 17 elementary particles, the last and most difficult of which were the 'Higgs Boson' which provides mass, and the 'Pentaquark', (both recently found/discovered by experiments at the Large Hadron Collider in Geneva).
>
> Unfortunately, because the model is unable to explain things such as gravity or the accelerating expansion of the Universe, the Standard Model is not a complete explanation. Yet M-Theory[22] **is** able to give some explanation of gravity and everything else, so it is believed by many to offer the only possibility of a complete explanation. What it lacks is any evidence.
>
> M-Theory basically suggests that sheets of force can form different shapes/structures which can hold energy, and their shapes may also cause them to react with each other in different ways.

So where does this leave us? If there are 3 possible core factors in existence, which might sit in isolation or be combined, (Matter/Energy, Force and Thought), there are 7 different ways that existence might be made-up, at high level, (however one of these is effectively a duplicate of another, so there are actually 6 core options).

With the background that we now have we can present these different explanations of our reality in a structured sequence:-

i. As mentioned above, the ultra scientific concept is that the core of existence is only made up of Force which is inseparable from Energy, and which creates what we perceive to be physical substance or Matter from which Thought later emerged as a feature.

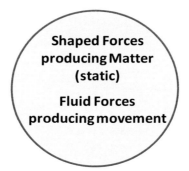

> Having removed the distinction between Force and Energy, there's no point in suggesting that Matter/Energy is a stand-alone option as it would amount to exactly the same idea.

As mentioned earlier, this removes one of the original 7 theoretical combinations.

ii.   The more traditional scientific possibility is that Matter/Energy provides substance while Forces act upon it to cause movement and sequence. In this context Thought and Life, would have to originate and exist purely as inevitable mechanical operations within Matter/Energy and Force.

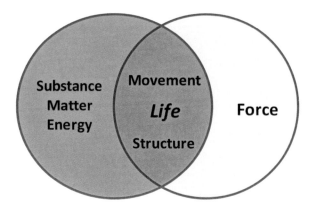

iii. We can construct a more 'open' scientific view which recognises that Life may require another type of stuff (Thought/Concept) that is not directly linked to Matter/Energy and because these factors both have static and dynamic forms, the presence or absence of the third element (Force) may be the thing that changes their state to generate movement and Time. This is shown in the diagram below.

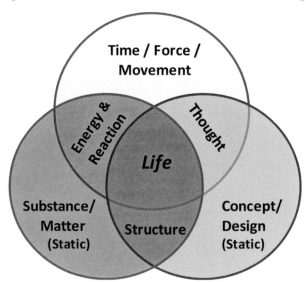

iv.   As we have seen, the view of Idealism is that Thought may explain everything. Unlike option 'i' this doesn't argue for the existence of a truly physical environment. Instead, it says that everything is in our minds and that physical reality is an illusion. However the philosophy also requires a degree of scientific backing and recently this has been found in the Dual Slit and EPR experiments which I cover later, (Chapters 6 & 7)

Collectively these factors imply that Thought is able to provide its own movement/sequence, but it doesn't have to provide physical power as we currently understand it.

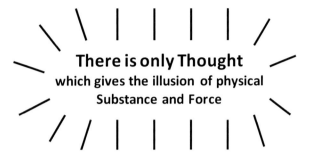

There is only Thought
which gives the illusion of physical
Substance and Force

Following the logic above, and for those that have not yet made the connection, option 'i' opens the way to another possibility.

v.   If the core of **_physical_** reality is Force, yet we also accept that Thought is fundamentally different to it, then the two core elements of reality are Force and Thought.

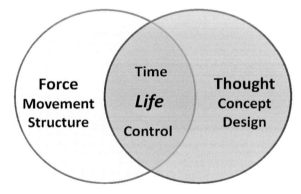

Force
Movement
Structure

Time
Life
Control

Thought
Concept
Design

vi.   The final option amongst the 7 theoretical combinations is Matter/Energy and Thought, however for it to be valid we have to decide where Force may exist between these two components. If force was contained entirely within Matter/Energy, Thought would be entirely an effect within the physical universe, which is no different to option 'ii'.

So in order to be distinct this option suggests that Thought contains an element of force, (although it allows for the possibility that both factors have their own elements of force). This is justified on the logic that within our bodies our Thoughts can be translated into physical movement by our brains, so it does seem to be possible for Thought to impart movement to physical matter.

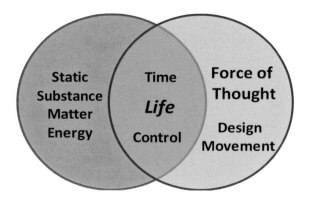

Static
Substance
Matter
Energy

Time
Life
Control

Force of
Thought

Design
Movement

The most likely scenario here is that both would have elements of force within them, however that would suggest that force is the underlying component!

So from the 6 viable combinations listed, which do you think might be correct? The bottom line is that we don't know, so it seems valid for each person to make up their own mind.

Yet even if we determine which of these factors form the base elements of our reality we cannot be sure which were created, and which were simply configured to produce their current form. However we begin to see that the raw things that may have emerged from the Big Bang will relate to this short list of options in various combinations.

5.2.1 Fields & Messenger Particles : the Force for Change?

> All matter originates and exists only by virtue of a force... We must assume
> behind this force the existence of a conscious and intelligent Mind. This
> Mind is the matrix of all matter.   (Max Planck - physicist)

All of the 4 basic forces in nature exert a field of influence which can be narrow, or broad.

As mentioned in Chapter 2, when we think about **_how_** a force pulls or pushes many people will initially imagine a <u>direct connection</u> between two objects, (like glue or string), so that if one thing moves the other will follow. Alternatively, where objects collide the point of contact would represent a very specific area where force was transmitted, creating a 'push'.

Yet as we have seen, there are many circumstances where that doesn't apply. As an example, Magnetism and Gravity both work across the emptiness of space... without a direct physical connection between objects other than a field of force – a field of influence. But what could a field be?

Generally a field generated by one object is a broad volume of space where other objects passing through that region are affected in various ways. We don't fully understand what fields are, but the latest theories suggest that they do two things[23]

1) They place a higher or lower amount of energy in their area of influence, and this will either give energy to, or draw energy from, objects which pass through them.

2) It is believed that fields can ***instruct*** objects which enter their area of influence to act or configure themselves differently.

If a field of force was like a sponge block with an unbroken network of fibres covering the region of influence it would be very difficult for things to pass through without damaging the structure of the field. As a result, science believes that a field represents a stream of miniscule particles fired off in all directions, but which follow a path back to the source.[23] Particles which hit an object as it passes through the field would either bounce off or be absorbed by the object, but once the object was gone the stream of particles would resume their original path as before.

Please remember that this is an idea not a fact.

As an extension of this idea, the tiny particles in a field are believed to act as **messengers** which either seem to deliver energy or, conceptually at least, provide a new set of instructions for the recipient object to act differently.

> As a good example of this, 'Photons' are the messenger particles for the force of electromagnetism which is why they also act as particles of light. Contact with them can cause an object to heat up, and then expand, etc..

Messenger particles have been identified for 3 out of 4 forces, but not for Gravity, (although models of existence often make provision for a 'Graviton' which might yet be found). So what are the ways in which such messenger particles could work? The basic choice seems to be that when a messenger particle bumps into or attaches to part of an atom, (an object), it either:

a) becomes part of the recipient object and therefore adds something physical to the original piece of matter, (eg. the energy of the messenger particle would increase the overall energy of the object),   or

b) somehow changes the object, perhaps making it more receptive to things in the environment, (eg. allowing it to absorb more energy from the environment), or causing greater attraction or repulsion,   or

c) communicates an instruction which tells the receiving object how to configure itself, (eg. to change size or shape, or vibrate harder and so warm up).

One factor which influences each theory is the knowledge that some of these effects are temporary, such as increasing heat, so if these messenger particles worked on the basis of options 'a' or 'b' above, they would have to leave the object at some stage taking those properties with them, allowing the object to return to its original state.

Yet if these effects were achieved by transferring information and *instructing* the object to do something new, ('c' above), it implies that the original set of instructions would get substituted or overwritten rather than building-up cumulatively, and this also gives us a problem if we expect the object to return to its original state when the messenger leaves, (as in the case of an object cooling down). For instance, it seems unlikely that if a messenger particle substituted an instruction, the recipient object would keep a copy of the original set of instructions just in case the messenger left!

We therefore see that messenger particles could work in different ways for different purposes, with some *substituting* details, while others work on an '*add-to*' basis.

In relation to option 'c' above, we have to realise that the ability to issue and act upon instructions would require potentially complex mechanisms to enable this to happen, that would either have to be built into the smallest particles of existence, or would have to form part of a universal 'Framework for Existence' that could impose changes from the outside. The added problem of communicating such instructions over large distances was once seen as a challenge to this notion until the experiments mentioned earlier revealed that instant communication over large distances was possible.

We will explore these ideas further in Chapter 8, however as a taster, if quite sophisticated mechanisms **are** present within the most fundamental particles of existence, we have to ask how they came about.

5.2.2   Induction

> All power corrupts, but we need the electricity.   (Anonymous)

The effect of 'Induction' has been known for some time, and is used as a prime way in which society generates electricity, however we don't really understand how it works.

Through this 'mechanism' energy can be made to suddenly appear in a place where it did not previously exist, without a direct physical connection, and for this reason it has attracted the interest of people who are seeking alternate ways in which the Big Bang may have been generated, (ie. producing a burst of energy that suddenly appears at a point in space).

> In basic terms when a magnetic field moves near an electrical circuit, (eg. copper wire), an electric current will appear in the wire even though the magnet doesn't make physical contact with the metal circuit. Neither do electrons necessarily jump between the two.

> Through an unknown process, the magnetic field is believed to transfer energy from the magnet specifically to the electrons which already exist in the wire, encouraging them to move around the wire circuit when they were previously tied to particular atoms. They will carry the additional energy with them, and the more energy that's transferred, the faster the electrons will move around the wire – making a stronger electric current.

> New electrons are **not** believed to be created because any new mass would use up vastly more energy than the electric current being generated.

> The current will later diminish either by resistance in the wire causing energy to be released, or by devices which tap into the current and remove some of the excess energy to produce light, heat, sound, etc.

Because we have to supply energy to the magnet, to get it to move in the first place, this process doesn't create energy out of nothing. In power stations that energy is generated by: burning coal or gas; through nuclear reactions; or by releasing water from dams. What we seem to be doing through the process of induction is changing the form of that energy by transferring it to the field and then to the wire.

Human beings do not generate electricity out of nowhere, (eg. through magic or pure creation). However induction is interesting because the energy appears within the wire of the circuit, not in the general environment/air which the magnetic field is passing through. It only 'materialises' <u>when the right conditions are established</u>. For instance, in broad terms, you wouldn't get any significant electric current being generated in your hand if you put it near a spinning magnet.

Most of the scenarios for the emergence of our Universe require something to exist before the Big Bang so could an effect similar to Induction, (but without coils of copper wire!), be the basis for a creation event, with or without God, (perhaps just involving the interaction of fields)?

Put another way, could something that pre-existed the Big Bang create the right circumstances for an Induction-type event to occur? Hold that thought.

6 Deeper Insights into the Nature of Physical Reality

Murphy's Law for 'String Theorists'

Anything in string theory that can theoretically go wrong will go wrong, but if nothing does go theoretically wrong then experimentally it is ruled out (Anon.)

This chapter begins our exploration of proven experimental results that directly challenge traditional scientific thinking.

Across the ages people have become increasingly aware that the secrets of our physical reality lie in the inner workings of atoms.

Because mankind has looked outwards to the stars and found that the broader cosmos appears to operate by the same rules which work here on Earth, we have grounds to believe that experiments *here* can reveal the secrets of physical existence everywhere.

The trick was to devise experiments, and develop the necessary technology, to probe the core of our physical reality, because at that level of existence we can't directly see what is going on. Yet that forces us into another layer of uncertainty as we then have to interpret the readings we obtain from our equipment - hence the quotation above.

In broad terms, sub-atomic particles can only be detected by recording the effects which those microscopic components generate on the wider world around them, eg.:

- where the sub-atomic component pushes other things out of the way, or
- our devices sense the radiation given off like a vapour trail along the path travelled, or
- particles collide and fragment, or
- the particle being monitored causes chemical changes when it encounters and reacts with other particles, or even detector devices.

Yet there are more subtle downsides to these detection techniques. As a start, there may be things that we cannot yet detect either because our equipment isn't sophisticated enough, or it isn't tuned correctly for a factor that we don't know anything about.

Equally, even though particles may only be visible for a tiny fraction of a second, there is still a delay between what happens and when it is recorded, which means that we may not know exactly where they *are*, only where they *were*, and the more accurately that we measure some aspects of their existence, (such as position), the less accurate other measurements become, (such as speed/momentum) because we have to apply energy and/or introduce other particles into the system to reveal those factors.

In short, we never get to see the full picture. This is the core of the 'Uncertainty Principle' in science: you can either see where sub-atomic particles are, or you can determine what condition they were in, or what they were doing, but you can't determine all aspects at the same time.

> There's a story that Heisenberg[24] (author of the Uncertainty Principle), was genuinely out for a drive when he was stopped by a traffic cop. The officer asked if he knew how fast he was travelling? Heisenberg replied: "No, but I know where I am".

A more recent joke works the principle even harder....

"Why are some Quantum Physicists so poor at sex? Because when they find the position, they can't find the momentum, and when they have the momentum, they can't find the position!" Sorry – couldn't resist it.

Despite this uncertainty, using various techniques, a huge amount of evidence has been gathered by science to show the types of component that exist within an atom, and how they might be assembled. This evidence consistently shows that each type of particle has its own set of distinguishing properties, which allows us to identify them with increasing ease and confidence.

Some of these extremely tiny components (such as 'fermions') do behave like true solid objects. Others do not.

Our knowledge only goes so deep, and the shakiness of our understanding is displayed when scientists admit that the very nature of some familiar *particles* is in doubt.

For example photons, (the individual component particles of light), and electrons, (the components of electricity), may actually be *spreading waves – not particles* or they may be able to swap between the two states. Some scientists even suggest that these components may have a degree of 'awareness' for what is happening around them.

These findings go straight to the heart of our subject about what our reality might be, but what led people to make these remarkable suggestions? Which of them might be true, if any, or could there be entirely different explanations which fall within our normal perception of how things behave?

The coming sections will talk about some simple experiments which were repeated in different ways because of the remarkable results they kept on producing. The sound of this may fill you with dread but I assure you that this **won't** be very technical, so please stay with it. A little bit of effort will bring its rewards!

6.1.1   The Dilemma of the Double Slits and the Nature of Fields

Simple things can have profound implications when they reveal that reality doesn't conform to established theory – ie. when things don't work out as they're supposed to.

The 'double slit' experiment, (first performed in the early 1800s), and its many variations, have shaken science to its very foundations over the question of what our reality actually represents. In part it also led to the establishment of Quantum Mechanics as a new scientific discipline.

The basic experiment is quite simple but it seems to break some of the fundamental rules of traditional science, providing us with a series of remarkable findings which should catch the imagination of all. **So even if you're not especially scientific, please stay with it!**

In the same way that mainstream science is wedded to 'cause and effect' it has historically been wedded to the idea that everything is based on particles. On this basis we have seen that science suggests all forces or fields of influence, (magnetic and others), are transmitted from a source to a recipient by tiny sub-atomic 'messenger particles' fired across a region of space.

These experiments cast doubt on what particles themselves may represent, but they also go much further, to say that those sub-atomic particles appear to :

- **physically change** when there's a 'choice' of routes to follow;
- **disguise or change their actions** when they are closely scrutinised;
- appear to **retrospectively change their destiny** based on what happens later!

Hard to believe but true – and this is something that theoretical physicists, (such as Albert Einstein, Stephen Hawking, and many others), have tried to rationalise without complete success. At headline level this and other experiments suggest that even deeper levels of existence & control may be in operation.

Formally, the basic demonstration is known as 'Young's Interference'[25] experiment and has been conducted using the tiniest sub-atomic particles of light (photons)[26] or electricity[26a] (electrons), as well as 'Carbon Bucky Balls'[20, 26b] which are groupings of **atoms** that are huge when compared to photons. It always produces the same dilemma:

- in some circumstances the particles act like discreet objects,
- in other circumstances they appear to be like 'fields of influence', spreading themselves across a volume of space, producing effects that resemble a wave.

It's worth describing the experiment as it forms such a crucial dilemma for science:

Imagine that you are directing a beam of light towards a screen with two narrow slots in it. If only one of the slots is open you would expect that any light hitting a wall on the other side of the screen would show an illuminated area shaped like the slot but larger, given the 'projector effect'. This is indeed what scientists find.

Now imagine that a second slot in the screen is opened-up. You would expect to see two bright images of the slots on the wall behind, but this **isn't** what scientists have repeatedly observed. Instead, they see a broad 'interference pattern', (per the image on the right below).

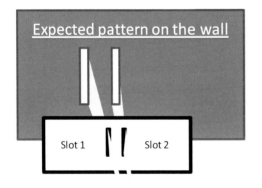

Expected pattern on the wall

Slot 1   Slot 2

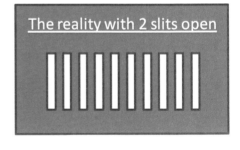

The reality with 2 slits open

Projector

With 2 slots open we see that the light is **spread out** and also forms a sharp pattern of **many light and dark patches** across the wall. The light is considered to be interfering with itself to produce the dark areas in a manner that is very reminiscent of a wave. The reason for this can be explained in two stages.

To explain, we need an analogy. Let's imagine that you have a bucket or pool of absolutely still water and you tap its surface with your finger. Ripples will spread out from that point, and in this way it's suggested that light may also spread out from the point where it comes out of a slot.

Secondly, we have to imagine that our smooth water surface is being tapped at two points a little way apart, generating 2 sets of ripples, (the equivalent of our 2 slots). Where the waves overlap they will do one of two things:

- if the waves are in sync, (a crest meeting a crest, or a dip meeting a dip), the size of the wave will 'double', whereas

- if the waves are out of sync they'll cancel each other out, removing any wave at that point.

A regular pattern of 'wave' or 'no wave' will emerge that stabilises after a short distance to form an **interference pattern** (per the emerging diagonal lines in the photos above). Those diagonals lines of 'no wave' can be equated to the dark patches on screen.

The dilemma in this experiment is that if a **single** slit was being hit by a wave it should still produce a spreading effect even if it doesn't produce an interference pattern. In other words, the 'block image' it would generate would be spread out to the same width as the interference pattern... but it isn't. The block image is a normal size.

An image that is **spread out** beyond the line of the projector beam _only_ seems to emerge when 2 slots are open.

We know from the 'redshift' theories in astronomy (chapter 4) why light is believed to travel in waves, so you may think that all of these images are entirely in keeping with the notion that beams of light travel in waves. But there are different types of wave and to ensure that everyone understands the difference between them, I hope you'll bear with me for a couple of paragraphs while I illustrate the point.

With one slit open we can draw a straight line from the projector to the edge of the slit, and continue that straight line to the edge of the 'block' image. The images are not fuzzy and there is no _spread_ of light beyond those straight lines. So if there _is_ a wave effect (per astronomy), it would be like a thin wavy line travelling in just one direction. Another analogy might be a guitar string vibrating in just one direction – say up and down, (holding it's line). In other words we get the impression that a single ray of light is **not** a **spreading** wave.

This is reinforced when we see that the beam of light from a _laser_ goes in a straight line, and the size of the dot it produces stays the same, regardless of how far the light travels before it hits something.

So it's <u>not</u> like a sound wave in air, or a wave in a mass of water, which are *ripple effects* produced in an existing **pool** of fluid when something else hits it, (eg. a boat causing waves on a still lake).

With *one* slit open photons or electrons behave as if they are something in their own right, travelling on their own thin course. They would be like a line of ships crossing the sea, bobbing up and down as they travelled in one direction over the waves.

However with 2 open slits the light appears to act like ***the body of water*** which is spreading ripples, and not behaving like a line of ships.

Having hopefully clarified the distinction, let's move on.

Recognising that natural or white light is a mix of different coloured light, (different wavelengths), scientists sought to eliminate any possibility that those different types of light might be having an effect on each other, so in later experiments they used a single wavelength of light produced by a laser. In this respect the results produced were typically red patterns of light on a white or black background. So in summary the different types of pattern which people have recorded show the following options :

No pattern – somehow the image of the slots/slits has been lost.

or

A clear 'block' image of one or other slot which can only be achieved when **light acts like a particle** going through one or other slit. This normally only happens when just one slot/slit is open.

The actual result achieved when a beam of light hits two open slots. This seems to indicate that the different parts of the beam interfere with each other to *spread out* the image while also creating an 'interference pattern' suggesting that **light is a field, travelling like a wave**, and not a particle.

Even more remarkable is the fact that the same patterns emerge when we repeat the experiment with ***individual particles*** of light (photons) or electricity[26] (electrons) rather than a beam, (a myriad of particles)!

With two slots open, *individual* photons hit the wall as a brief point of light, but if we map where a lot of these individual photons land on the wall, they also produce a broad interference pattern. Yet none of these photons will pass down the apparatus at the same time. Each one will die before the next one is issued, so they don't have the opportunity to co-ordinate their actions, but the **interference patterns do show that there is a level of co-ordination**.

Images courtesy of Dr.A.Tonomura from his experiments in 1989.

But with 1 slot open the dots still build into a single bar or 'block' image.

So let's think about this a bit more.

As we believe that particles are issued by a laser, and a dot is recorded at the end, (suggesting that particles are detected), most people try to imagine what those particles might do as they travel down the apparatus.

With one slot open it's assumed that the photons or electrons remain as particles throughout, to produce a sharp block image. The interest lies in what may be happening with 2 slots open. On the assumption that there is no external influence (ie. beyond the particles themselves), two main options were initially proposed; either:

a)   The photons/electrons **transform themselves into a spreading wave** and then collapse back into a single dot/particle when they hit the screen or detector.

b)   They remain as particles but **adjust their trajectory** which causes them to take their place within a distinct pattern that seemingly anticipates what other particles might contribute-to when they come into existence later.

In both of these cases the particles must 'know' whether 1 or 2 slots are open in order to adjust their behaviour, and this has led to suggestions that the particles have a crude level of 'awareness', because the screen **doesn't** emit any signal. While I realise that awareness is an emotive word it is hard to think of another in this context. Particles operating in isolation, (without external influences), would have to reach out, detect, and respond in different ways to their circumstances.

> It has been suggested that if each particle transforms into a wave before it reaches the screen, it would need no 'knowledge' nor any special powers of detection, and could simply encounter the slots to know how many were open... but this misses the point. If a simple wave encountered only one open slit, then it would instantly have to adopt particle form to avoid any spread. In other words, it must still behave differently than it would when 2 slots are open.

> A degree of crude awareness is difficult to avoid if people try to explain what is happening in these ways. On the other hand, if you can't accept that particles have awareness you must find other solutions... which has proved rather difficult

Yet the problems with options 'a' & 'b' above don't stop there.

If particles do transform into a broad wave then this might account for the spread and pattern of the dots on the screen, but it doesn't explain why the wave would crystallise into a dot at one particular place along its length rather than any other position, **and** still fall within an overall pattern that is co-ordinated with other particles which are either already dead or haven't been issued yet.

On the other hand, if the photons/electrons *stay as particles* and somehow determine that 2 slots are open before adjusting their own trajectory, they would be displaying an ability to steer their own course that has never been seen in any other context. There would also be no explanation of how they could generate an interference pattern, because they weren't waves. In the absence of any other influence to guide them, suggestions that they could steer themselves would imply that these particles had even greater levels of awareness and responsiveness, which is why this seems even less plausible to many people.

You can see why scientists are keen to find other explanations which are less challenging, but unfortunately while there is one possibility, (that I will talk about later), it's not being actively pursued because there is no direct evidence for its key mechanism. Scientists still seem to focus on solutions based on the things they can detect.

Yet in broader terms, there are glimpses of normality because interference patterns can be predicted, which suggests that they arise because the photons/electrons act in an inevitable way. That is the hope of many people.

As none of the proposed solutions were particularly palatable options, the obvious thing to do was to carefully monitor whether each photon was going through one particular slit (as a particle), or was spreading out like a wave across both of the open slits.

This is where we come to yet another remarkable finding…

When monitoring equipment is actually used to closely observe what is happening at the slots when they are both open, the interference pattern **disappears** to only leave a 'block image'.[27] It almost seems as if nature doesn't want us to know what's happening.

Yet the interference pattern returns when the monitoring equipment **isn't used, (even if the apparatus is left in place**)! Intelligent particles – or a response to our Thoughts, as some have suggested?

Observation, (ie. 'seeing'), is an action that *receives* signals, it doesn't *issue* them, so it shouldn't be able to cast any influence over anything. The obvious possibility was that the monitoring equipment itself was causing the effect, so scientists initially suggested that either :-

- electromagnetic fields *generated by the monitoring equipment* may have been having an effect on photon waves causing them to become particles, or
- the closeness of the monitoring equipment to the slits blocked/disrupted the waves, causing them to return to a particle state.

While some of the early equipment was indeed found to be generating some fields, and some devices did get so close that they may have been in the way of the particles, the designs for later experiments compensated for this, but the results didn't change[27a].

In some quarters this reinforced suggestions of crude awareness, or a response to external influences such as the thoughts of an observer. Others suspected that because photons & electrons cannot normally be seen 'in flight', the disappearing patterns must in some way result from the techniques being used to make detections.

Experiments conducted in 1987 and 2012 claimed to have obtained 'which path' information by monitoring the slits *'without removing all'* of the interference pattern [27b] but these were **not** conducted in exactly the same manner as the original, which has led some to challenge the conclusions.

Equally, I want to be clear that **I have not seen references to any scientific paper** which supports claims on the web that interference patterns only disappear *when a person looks through the equipment*. We cannot see these things in real time - the action is too fast. So despite the persistence of these rumours, the suggestions are **not** fact, and at the very least are highly disputed.

However it was still necessary to explain how photons and electrons could change their behaviour in the basic circumstances of the experiment if there was no physical influence, and if the only difference lay in the path taken through the apparatus a suggestion began to emerge that particles may naturally produce an interference pattern, (some sort of manifestation of the image), where there was uncertainty about which slot was used, and that monitoring the slots would remove that uncertainty.

As strange as this sounds, scientists would not be risking their reputations on such speculation if the verified experimental results were not so startling. But they are.

Incidentally, it could **not** be suggested that the human mind *causes* the interference pattern because Thomas Young never expected such an effect to be generated in the first place.

Scientists needed more information about how the particles were operating, so they devised a series of tests to try and trick nature into revealing what was happening.

We now reach the cutting edge of scientific research where there is intense debate over whether 'awareness' or 'thoughtful observation' causes a direct change in physical matter. Due to its implications scientists tread very carefully in this area, so it is important to clearly understand the nature of key experiments in order to make up your own mind about what they represent.

6.2 Wheeler's Delayed Choice and the Quantum Eraser Findings

John Archibald Wheeler first proposed his Delayed Choice Experiments in 1978 (*Mathematical Foundations of Quantum Theory*). A number of different versions test whether single photons in the dual slit experiment can sense their environment and adjust to it, or remain in an indeterminate state until detected.

He proposed ways in which the photons could be monitored without altering the set-up of detection equipment, while being able to change both the number of open slots and also which one was being monitored *after* the photons had been fired at the slots. This included the choice of whether to detect or not.

It took a while for technology to catch up with theory so the first actual experiment was conducted in 1984 by Alley, Wickes, & Jakubowicz at the University of Maryland Surprisingly to many, it **did** provide more evidence that observation may be influencing the outcome. Results were presented at the *Proceedings of the 2nd International Symposium: Foundations of Quantum Mechanics in the Light of New Technology*. Tokyo, Physical Society of Japan, 1987.

Wheeler himself had clear views about what was happening – he said : "Actually, quantum phenomena [the photons and electrons] are neither waves nor particles but are intrinsically undefined until the moment they are measured." (Scientific American, July 1992, p. 75). **Put another way, he believed that sub-atomic components didn't take any form until they were *observed/measured*, in which case they would either adopt the characteristics of a 'particle' or a 'wave' depending on circumstances.**

Since then, different versions of Wheeler's experiment have been devised, and the Quantum Eraser version of these experiments has arguably been the most influential. It was first performed in 1999 by Kim, Yulik, Shih, & Skully and produced a storm of debate. The set-up of the apparatus is shown in the diagram below. It looks a little daunting at first sight but in fact it's quite simple, so please stay with it. Again - a little bit of effort to understand the experiment will give you some cutting edge insights!

This experiment has sophisticated detectors instead of a back wall, and it also uses a proven principle in science – that a single photon of light can be split in half, to create two 'entangled paired particles' which **exactly reflect each other's state of being at any point in time, even if they are miles apart**[19]. In other words, to a degree, you can see what is happening to one of the split particles in the pair, by monitoring the other.

With both slits in the screen open (A & B), an original single photon will be allowed to pass through one or both slots as normal and whatever emerges from each slot, (wave or particle), is then split into two and the resulting 'paired particles' or 'split waves' then travel through different sections of the apparatus, (see diagram below).

> Photons coming out of slot A would follow the course of the red lines on the diagram, while those from slot B would follow the course shown by the green lines. (The actual colour of the light wouldn't be affected).

> If waves emerged from both slots then two sets of paired particles would be created by the splitter.

> One of each pair will immediately be sent to detector D0 while its partner will follow the longer route along either the red or green course.

A simplified version of the equipment set-up can be presented as:

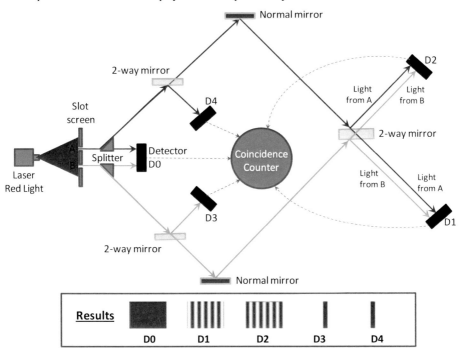

There are 8 possible paths through the apparatus (shown by the lines) with 5 end points that each have a detector to capture the patterns produced, (D0 – D4).

The pale blue oblongs represent two-way mirrors where half of the photons bounce off while half of the photons pass-through. The ordinary mirrors, (shown as dark blue oblongs), make <u>all</u> photons change direction.

As mentioned before, the first of the paired particles, (sometimes known as the 'signal' photon), should always reflect the original state of each photon as it emerged from slots A & B and this will always be captured at detector 'D0'. The 2nd of the paired particles, (known as the 'Idler Photon'), will travel to one of the other detectors (D1-D4), and if the image pattern recorded at any of these differs to D0 then we know that the change of state, (betwcen particle and wave), occurred somewhere after the splitter.

If you look at the paths that the Idler Photon (2nd of the paired particles) might take, you will see that detector D4 can *only* take light that comes from slot A, while D3 can *only* take light from slot B. These detectors have only ever shown the 'Block Images' of *particles*.

The light hitting detectors D1 and D2 can receive light from both slots and interestingly these have **always** demonstrated an interference pattern, characteristic of a *spreading wave*.

The results are quite distinct for each detector, and will be built-up as a series of dots per the black & white dot images above – but using red light.

If waves hit D1 & D2, the photons must either be changing their state/form after they pass the first of the two way mirrors, or they must be waves from the very start.

If interference was caused at the slits we could expect interference patters at D3 and D4 but we don't get them (not even an overlap of block and interference images). Conversely if the waves were kept separate until the tail end of the apparatus, we would expect to get a *spread-out* block image at D3 and D4 – but we don't.

So if they're not waves throughout, they must start out as particles and then change after they get past D3 or D4. The problem with this is that if the photons/electrons are particles, they can only take one path, so there is nothing to cause an interference pattern at the tail end even if they do change into a wave.

In broad terms, this is why physicists suspect that photons/electrons are formless until detected, when they will adopt characteristics relevant to their circumstances... which to me again suggests a degree of awareness.

Once again people place emphasis on the fact that if the photons/electrons **are the only active elements in the experiment,** the only distinguishing factor must lie in the knowledge about 'which path' was taken – through slot A or B. Where there is clarity we always get a tight block image, but where there is uncertainty we always get a spread-out interference pattern.

Yet these notions are entirely dependent on assumptions being made about factors we can't detect, and therefore many people will remain sceptical about this line of thinking.

Before going into a deeper analysis and a possible solution, I need to tell you about one more startling finding from these experiments.

The image at D0 is reported to show no pattern at all, and people therefore suspected that this was because it represented an amalgamation of all the different results from the other detectors, (ie. the different images recorded at each of the other detectors would overlap so completely as to remove any pattern).

My concern about this claim is that if block images were overlapping the interference patterns they would *either*

- remove stripes in the middle of the detector but leave stripes on the outer edges, or
- if the interference patterns at D3 and D4 were slightly offset, they would show a very even red background with two brighter block images overlaid

- rather than no image at all.

I haven't been able to get clarity on this point so we have to go with the findings as reported – including those below.

As a result of this debate, and as part of the equipment set-up, scientists analysed the hits recorded by each pair of particles at the 'Coincidence Counter', and they found that the pattern of hits recorded at D0 were exactly the same as those recorded by each of the other detectors. The patterns from the different detectors <u>did</u> seem to be overlaid at D0 - however this 'exact match' between the detectors posed a difficult question.

The path taken by the 'Signal Photon' to D0 is a lot shorter than the paths taken by the 'Idler Photon' to the other detectors, which means that the first of the paired particles to end its life will always be the 'Signal photon' at D0 and yet it always seems to know the final state of its twin which 'died' later.

The significance of this lies in the fact that the fate of the Idler photon is uncertain, as it may or may not pass through the 2-way mirrors, and so we don't know which detector it would hit or which image it will contribute to. For instance, if a photon was originally destined to produce a spread out interference pattern, but it then bounced off the first 2-way mirror, it would be changed to a narrow block image pattern?

So how could the Signal photon predict the outcome for the Idler photon with 100% accuracy? Most people concluded that it couldn't under the rules of cause & effect, therefore the Idler photon must somehow, retrospectively, affect the Signal photon. While the paired particles are known to retain a direct influence on each other while they both exist, it was <u>never</u> envisaged that they could retain that link after one had died – implying they could ***go back in time to change the earlier result***!!

I discuss the theoretical implications of time travel in another book and to cut a long story short, it is not considered possible to go back in time, although it may be possible to go forwards faster than normal. So this finding was highly controversial – not only because it would mean altering the position of the signal photon after it had died, but also because it meant undoing the results which had been recorded at D0 and stored separately by the apparatus!

A way had to be found to explain things without the 'time travel' implications, and I am only aware of 4 suggestions (which include the possibility of a 'hidden pool' below):

- The end state of both particles can somehow be known in advance, (predicted), or
- Both particles undergo the same influences to result in the same outcome, ***however*** there are various ways in which they could differ, (discussed later).
- The suggestion by Thomas Campbell[28] that our existence is shaped by reference to an information layer within reality, (discussed in a moment).
- That the Thoughts of an observer shape the outcome that is measured/interpreted.

Campbell's basic idea is that every particle in existence is defined/configured/shaped by data that is held about it within another dimension of existence that we can't see or detect. (Across 800 pages of his book he identifies many experimental findings to support his ideas, but most commentators that I have come across remain highly sceptical). His explanation for the Quantum Eraser findings works like this:

> When a particle is split the separate parts create two linked records in the 'information layer' of existence. The end state of the first 'signal' photon is recorded as it dies at D0, and then **both** records are updated

when the second 'idler' photon dies at one of the other detectors some time later. In this way the two records are brought into line.

When other parts of reality, (such as the 'coincidence counter' above), separately record what happened they might also link back to the information layer. So when the results on the coincidence counter are inspected they will reflect the state of the records in the information layer, rather than what actually occurred in sequence.

Reality would seem to be retrospectively adjusted, when in fact the process relates to changing *data* within the normal flow/sequence of events in Time.

It's an interesting idea but it lacks direct evidence to prove the existence of this information layer, and while it strengthens the case for data/messaging etc. it doesn't necessarily imply Thought. As we all know, data can be captured mechanically without any Thought being applied.

The other obvious challenge to this way of thinking comes from the need to rationalise how a separate recording device would somehow be able to link to the particle records.

On the other hand, this concept doesn't remove the possibility of a Thought-based influence either, and the presence of a data layer may even enhance the chances.

6.2.1   A Consideration of Viable Explanations

The Double Slit experiment and its many variants are remarkable because they demonstrate highly unusual behaviour.

Depending on whether you feel that the particles are acting in isolation, or are receiving external influences will lead you to different conclusions about what is happening at a level of existence that we can't see directly.

Most **Materialist** explanations have been based on the things that were detected which indicate that the photons are the only active factors in the experiments and this has led to a small number of suggestions which imply that photons have the capability to:

- sense their environment and then change their behaviour accordingly; and/or
- change their state of being from particles into spreading waves and then back again, (never considered in any other circumstance); and/or
- change course on the basis of information received.

As these directly challenge long-held ideas about what the physical world is, you can hopefully see why these ideas are still treated with scepticism.

From a **Dualist** perspective, the symptoms being recorded provide clues that other types of stuff may be involved, (which we can't detect): from the nature of spreading waves; to the need for crude awareness under certain circumstances; and even reactions to close scrutiny/observation.

Finally, there are ideas which closely resemble **Idealist** thinking which claim that photons have no shape or form until they encounter a situation which requires an end outcome, (such as measurement or observation), which forces them to crystallise as a dot in a location appropriate to the circumstances.

By having no definite form, photons could exploit the properties of both a particle and a wave as appropriate to the circumstances - which seems to strengthen ideas that

- data/ information may be shaping outcomes;
- imply a degree of awareness; and
- in relation to the 'D0 dilemma', may offer solutions that avoid Time travel.

The surprise to me is that the Materialist camp hasn't tried to search for additional types of *tangible* influence. Despite clues that the only way to generate a spreading wave is by causing 'ripples in a pond', the assumption seems to be that our equipment would have detected such things by now, and the fact that we haven't means such things can't exist. Yet we haven't detected 'Dark Energy' either. Could that be the pool – possibly in association with background radiation? It's easy to see how the speculation might grow.

However in abstract terms we still need to consider if a hidden pool could explain the results presented above, and I will do this in the section below.

For the moment, with so many strange results and so many overlapping factors it is very important that we carefully separate the different elements to prevent them from being confused, and as all three of the main philosophies have arguments to explain the experimental results, there's a need to return to basics.

As a starting point, I feel it is significant that the **predictability** of an interference pattern strongly suggests that **it arises from inevitable and probably 'mechanical' factors**. The results of known experiments never change. Even a disappearing interference pattern can be predicted when we set-up monitoring equipment at the slits.

Secondly, we have always recognised photons and electrons as discreet 'packages' not as spreading waves and this is how we issue them from a laser. At the end of the experiment they are always detected as dots, not as faint impressions of the full image. To many this will reinforce the idea that they start and end as particles.

Equally, there are very few ways to generate an interference pattern; the prime one being through an overlap of two spreading waves, and because the interference pattern only emerges when two slits are open it's not unreasonable to suppose that those overlapping waves will originate from the open slots. Yet there is a suggestion, (mainly from the Idealist camp), that an interference pattern is the natural **manifestation** of uncertainty.

If we stay with physical causes, the question is whether the waves come from the particles themselves, or something else – such as other types of stuff?

However if the particles are the only active elements in the experiments, we have seen three main ways to explain how the block of interference images could be obtained:-

a) The photons/electrons might be physically changing into a spreading wave en-route and then back into a particle before they were detected.

b) The photons/electrons were always waves (we just mistook them for particles), but they still crystallised into a dot at the end.

c) The photons/electrons lacked any shape/form until they were required to adopt a sensible/relevant format when detected.

Diagrammatically the underlying three states of being can be shown as :-

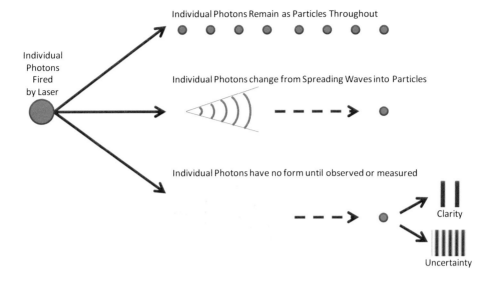

The clearest indications that a change of state/form **may not** be right are because

   i.   the same experimental results can be obtained if carbon bucky balls are used instead of photons and electrons, and those definitely don't change shape.

   ii.   in the 'real world' *spreading* waves only ever arise when an object such as a particle, boat, or train causes ripples in a pool of **other** stuff.

Other more detailed challenges can occur along the following lines:

   iii.   Something has to cause the particles to produce different effects, and if that is not a direct physical influence from something else, the particles themselves would seem to require a degree of awareness to recognise and react to the differences in their circumstances.

   iv.   There is no rationale to explain why a spreading wave, or a formless region of influence, should crystallise into a dot at one particular place within its broad area, or why this should fit into the broader pattern being generated.

You may also want to consider the physical variables that we know are present when photons & electrons pass down the Quantum Eraser apparatus. There are 3 main ones:-

   a)   whether one or both slots were open,

   b)   whether a detector could receive photons/waves from one or both slots,

   c)   whether a photon would pass through a 2-way mirror or bounce off it?

What influence could any of these things exert?

There is an assumption that 2-way mirrors won't affect the nature of a photon or electron as they pass through them, but that is questionable.

If waves fragment every time they go to through a slot, and a 2-way mirror is effectively a reflective surface with lots of holes/slots in it, they might be fragmented by every one of those tiny holes causing a myriad of additional overlapping waves. In such circumstances any interference pattern would either be broken down into very thin bands, or might even disappear altogether. Is that significant?

Alternatively we might argue that the glass in a 2-way mirror would prevent any wave from getting past it, and only particles would be able to pass through. Without knowing the stuff involved, it is hard to say.

Partly as a way to resolve some of these issues, it would be useful to know how many dots were recorded by the detectors, compared to the number of photons issued by the laser during the experiment. Unfortunately I can't find anything to tell me. At the very least we would expect twice as many, due to the particle splitter, but different theories produce different multiples, especially if every wave is being fragmented and each fragment should crystallise into a dot.

It would also be useful to clarify whether it was just the general patterns achieved at D0 that matched the results at other detectors, or whether the position of every signal photon recorded at D0 matched the location recorded by the associated Idler photons detected elsewhere. The working assumption by many analysts is that every single dot was in exactly the same position.

Interestingly, the 'Hidden Pools' suggestion **can** explain all of the **primary** results from the experiments in a Materialist context without resorting to special properties for photons & electrons (especially when those couldn't apply to carbon bucky balls). The findings which a hidden pool **can't** provide any apparent explanation for, are :

- why an interference pattern disappears when the slots are being closely observed.

- how the results at D0 can exactly match the results at the other detectors in the Quantum Eraser experiments.

Yet these are problems for other theories too, and a hidden pool may only be part of the story.

The 'Hidden Pools' Explanation for Basic Results

This approach *always* regards photons, electrons, and carbon bucky balls as particles. Like ships on the sea they would follow a straight course, riding the waves they were creating in a hidden pool of other stuff, (their course appearing to act like a wavy line).

However if two slots are open then the ripples in the hidden pool would generate an interference pattern and this would affect the course taken by the particles. The diagonal troughs of an interference pattern would become a point of resistance to the normal path of 'ships over the waves', causing the particles to change course.

Some particles could simply find themselves on a slightly different heading, but others (possibly a majority) may be 'captured' by the gulley of the interference lines, (the steep sides of the waves effectively falling along the 'side of the ship' compared to the direction of travel). This could lead them to follow the exact line of those interference patterns, explaining why there are concentrations of dots in some places on the detector, (aligned with the interference troughs), but still some dots appearing in darker areas (aligned with the normal waves).

So how could this explain the results of the Quantum Eraser experiments? For ease, let me present the apparatus layout again.

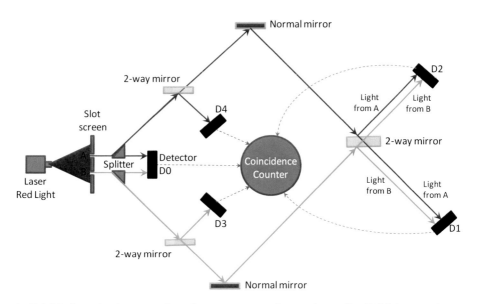

At D1&2 there is always an interference pattern, but at the earlier D3&4 we only ever get block images. The explanation for this would probably lie in the fact that **this hypothesis separates particles and waves yet probably allows both of these factors to travel down the apparatus together**.

Waves in the hidden pool would always go through **both** open slots, but a single particle could only ever pass through **one** slot. Whether that was A or B, the first thing that an Idler Photon would reach is a 2-way mirror. If it was deflected at that point and hit either detector D3 or D4 it would produce a block image because it would have been riding a **single** wave up to that point.

Yet if the particle and its wave weren't deflected and passed through the second 2-way mirror they would **then** encounter the wave from the other slot and this would cause an interference pattern after the 2nd 2-way mirror, because the waves would be perfectly timed to arrive together and would therefore overlap. In this way only interference patterns could be recorded at D1 & D2.

The challenge for this theory is in explaining how the results at D0 will always match the other detectors D1-4. While there may be configurations of the equipment that could explain most of the detector results, (see below), I can't think of any which could explain them all. For instance :-

> If the beam **splitters** prevented the spreading wave from slot A overlapping with the spreading wave generated at slot B at D0, then the explanation above would hold true for D1-4, **but** there would also be no way in which an interference pattern could be generated at D0 without such an overlap.

> It would be necessary to show that overlapping waves could occur in the channel up to D0 (in order to achieve an interference pattern) but also be prevented from overlapping on the path to D3 & D4, (to preserve the main results). Yet such an arrangement would **always** produce an interference pattern at D0 and could never explain how a block image could be recorded.

In the absence of a complete explanation it may be that we need some other mechanism like Thomas Campbell's suggestion to keep all the D0 results aligned.

In relation to the disappearance of interference patterns when people try to closely monitor what is happening at slot A or slot B, it might be suggested that equipment could be blocking the path of the hidden waves from the pool, rather than blocking the particles themselves. Yet this doesn't really equate with the results as they are being reported – ie. the equipment can be in place while there is an interference pattern, but this disappears when scientists try to use it. If this is the case, (as most articles I have read seem to suggest), then there is a strong pointer to an ***additional*** influence – which some people may equate to information/awareness or even Thought.

I suspect that many people would probably remain sceptical about suggestions of awareness in particles unless there was additional evidence, however those suggestions **were** strengthened by two other experiments that I will cover in Chapter 7.

Early reports about tests based on 'Bell's Theorem' since the 1980s were somewhat exaggerated, leading to myths that that nature was based on a devious consciousness which operated at the deepest levels of existence. It would be fairer to say that these experiments were effectively designed to kill off all suggestions that broke traditional Laws of Physics - **but they produced the 'wrong' result**!

You will need to make up your own mind about the findings in the next chapter, but my personal feeling is that they provide evidence which supports a different set of rules rather than randomness and consciousness.

As you make your own assessment of these issues, can I suggest that you begin by deciding your initial position on the following 5 questions, which I feel are the key to unlocking the mystery. You can then test your perspective accordingly.

1) Do you believe that photons/electrons are the only active elements in these experiments; or that there are other things involved?

2) If there are other things, what could they be?

3) Is it necessary for photons/electrons/carbon bucky balls, to **detect** when 2 slots are open instead of one, or just to react to direct physical contacts? In other words, do they need a degree of awareness?

4) Do photons/electrons transform themselves as they pass down the apparatus?

5) How can spreading waves crystallise into a dot whose position also conforms to a broader pattern with other things that didn't exist at the time?

In the spirit of keeping an open mind, let's also bear in mind the range of possibilities for generating an Interference Pattern. People have become fixated on one particular mechanism, the **physical wave**, but there are other possibilities:

- Magnetic or other types of field - (not detected).
- Physical contact (things bouncing off, or reacting with, each other – eg. the sides of the slits)
- Filtering, (which could potentially be introduced in some experiments through polarisation, etc.) and
- Manifestations (ie. if things appear, they will only do so in a certain way)

When I considered these factors I found that they were each specific to one type of apparatus, if they applied at all. They were also hard to apply across all scenarios concerning photons, electrons, and carbon bucky balls.

We each have to decide whether a single solution is more likely for the Double Slit experiments than separate explanations for those 3 types of particle.

7 A Rigged Game, Instant Control, & Idealism

The double slit experiments shook science to its core by revealing that at the tiniest levels of existence particles can sometimes act like objects while at other times they can act like wave fields. Even worse, they seem to change their state when people try to closely monitor what they are doing. The actual results are bizarre, but true, yet the interpretation of what they meant has polarised scientific opinion. More was to come.

7.1.1   Instant Control over Large Distances

We saw in the last chapter that science was making regular use of a phenomenon known as 'paired particles', where a single photon could be split in two to create an 'entangled' pair in which the state of one particle, (eg, spin, polarity, etc), would always be reflected in the other particle, even if they were separated by large distances, so anything you did to one would be <u>instantly</u> mirrored by the other.

> In fact the meaning of mirroring in this case is quite literal, because the spin/polarisation etc. of one particle will be the exact reverse of the other.

These properties came to light, (if you pardon the pun), in the early 1930s and were the subject of heated debate between Albert Einstein (a proponent of traditional physics), and Ernst Schrodinger (an advocate of the emerging discipline of Quantum Mechanics). In broad terms, Einstein said that it wasn't possible for anything to communicate that quickly as it would break the speed of light, whereas Schrodinger said it was really happening. But who was right?

The first robust evidence that Schrodinger was right came from experiments conducted by Alain Aspect[31] in 1981/2 - however these experiments have other connotations linked to the presence or absence of randomness/control at the deepest levels of reality, and they will be discussed in the next section. I therefore want to clear the ground to give a better illustration of the 'faster than light' issue.

An experiment conducted by Nicholas Gisin[19] et al. in 1997 at Geneva University sent two entangled light particles in opposite directions along fibre-optic cables, each approx 6 miles long (10km) where one was then subjected to a spin before it died, and the other was left alone and yet it instantly mirrored the change in spin at the other end of the apparatus. Nothing should have been able to communicate between the two, as they were moving away from each other at twice the speed of light, yet the result indicated that by influencing one particle towards the end of its journey the communication would have to travel **at least** 10,000 times the speed of light in order to affect the other particle before it died.

It has been a direct challenge to established principles in physics that nothing should be able to travel faster than the speed of light, so various possible objections to this conclusion were floated within the scientific community. Those who don't like the implications will throw lots of maths at you to describe why these effects aren't really real, but to ordinary people they are – absolutely.

It was suggested for instance, that while the results did point to a hidden connection it wouldn't violate the speed of light because the particles were never that far apart – they travelled down wound up cables in the car park which were then returned to different detectors in the same lab. However I am reliably informed that the detections were actually made between the villages of Bernex and Bellevue which lie on opposite sides of Geneva.

Recent experiments, (2015), at Delft University of Technology also claim to have eliminated all of these challenges and proven that instant communications are indeed happening.[32]

To make the points again:-

- The particles have no known **physical** connection.
- In all practical respects they can communicate instantly even if they are miles apart and travelling away from each other at twice the speed of light.
- That communication may **not** transfer physical material, but it **can** generate physical effects – eg. a change of spin in the recipient particle.

Clearly these proven findings break the perceived Laws of Physics at our level of existence, and the connection which these photons retain must be equivalent to a crude sense of awareness of each other; an intangible link related to communication and possibly information.

In the earlier Quantum Eraser experiments we also saw that this effect appeared to go one step further to produce a really strange result: that the 'signal' particle which was the first to 'die', always seemed to correctly predict the final state of the other 'idler' photon, even though that second photon might be changed/influenced at a later point in time.

An attempt to explain this by Thomas Campbell[28] suggested that every particle in existence was instructed to take shape by data held about it in an *information layer* (or additional dimension of existence), and paired particles would be linked via this information layer rather than some physical link within our 3 dimensions.

To traditionalists these are very awkward facts, and ones that aren't going away because the experimental results have been repeated. They show that nature has an ability which contravenes **one** of the perceived Laws of Physics at our level of existence, (speed of light). The other Laws are largely untouched, however it does also challenge the principle of **physical** causality.

These effects point to a capability, but they don't say what allows these actions to happen. Whether these findings point to another dimension of reality, or the presence of a 'controlling field', or a 'framework for existence' (Chapter 8), or indeed any other factor - has yet to be established.

However in terms of *using* the effect, people do see the beginnings of an explanation for Thought. The downside to this notion is that people don't believe that our brains work on the basis of paired photon or electron particles.... but in terms of electrons in particular, who knows what may be revealed in future?

Yet from a broader philosophical viewpoint, it does seem to provide surprisingly strong support for the theories of Idealism, including explanations of the double slit experiments which are very close to this way of perceiving existence, (such as wave/particle duality).

Your view of whether or not physical matter is real, (ie. has true substance), will then lead you to decide whether the effects are more symptomatic of Dualism.

7.1.2   The EPR Paradox – and Bell's Theorem

Traditional science, (defended by Albert Einstein, Boris Podolsky, and Nathan Rosen in the 1930s - 'EPR'), held that nature would always follow the traditional Laws of Physics and the principle of physical Causality, so if an experiment showed a particular result then its outcome would effectively be pre-set by local factors in the environment that led up to it.

Any perceived change that seemed to result from an act of simple observation would be nothing to do with the observation itself but due to some other pre-determined factor... or hidden variable. In other words, cause and effect would be maintained within physical matter, (without any other influence such as 'awareness'/Thought).

Their starting point in relation to the Double Slit experiments was to argue that the process of creating photons or electrons, (eg. within the laser), would determine their original form and that the *circumstances* of the apparatus which they travelled-through later might change them. A different version of this notion is that the particles do not change but are influenced by a 'hidden pool' of stuff that surrounds us all.

The alternate proposition from the advocates of Quantum Mechanics (such as Ernst Schrodinger, Niels Bohr, and Werner Heisenberg[24]), suggests that nature, at the sub-atomic level of existence, **may exist without any shape or form,** (wave or particle), **until nature requires an end outcome**. That end outcome could be when it is detected/observed/measured.

Einstein said this was like believing that the moon only exists when you are looking at it, yet he couldn't deny the evidence of the dual slit experiments, and couldn't explain how they might work on a 'cause and effect' basis without breaking other fundamental laws of existence that he had helped to formulate! (As an example : explanations which require particles to reach out and sense their environment, (using unknown mechanisms), would have to operate faster than the speed of light – which he said was impossible). Something was wrong somewhere.

For many decades there was no apparent way to test which of these philosophical ideas was correct. The public, and indeed many traditional scientists, were un-persuaded that their solid reality was only taking shape when observed, however in all practical respects, the new science of Quantum Mechanics was delivering good results. It worked, so in their day jobs scientists just got on with it.

The philosophical questions remained in limbo until John Bell[30] (b.1928 in Belfast), provided a way that might demonstrate which of the two ideas about existence was real. This effectively recognised that we didn't know all of the factors at play, but he used the principle that if there was an even chance of something pre-determined occurring, (as a result of cause and effect), then a series of experiments would ultimately show that you could guess what was happening and be right 50% of the time, however if the 'game was rigged' there would be a significant distortion in the results. This would occur if certain outcomes were not based on local cause and effect but subject to seemingly unrelated last minute factors - the QM view of the world.

Bell's solution went further to try and 'level the playing field' because he recognised that if nature was trying to cheat by effectively 'changing the result' if we 'dared to look closely at what was happening', the results of the experiments couldn't be rigged if the way in which each play was to be made was committed-to **before** we made the prediction and we then observed what happened, (ie. if we got nature to put its cards on the table before we guessed, and then we checked the results)[3].

This description is an analogy, and his formula is rather more complex than that, but after considerable scrutiny it is regarded as a solid concept. So in principle, any significant distortion in results away from our notional 50:50 would either disprove strict causality or point to additional influences.

If we continued the analogy of a card game we might guess red or black[3], but in reality to reflect the 50:50 option Bell suggested measuring spin – ie. up or down / clockwise vs anticlockwise. Put another way, it is **not** a replica of the double slit experiment which would measure a block vs interference pattern, but it <u>would</u> illustrate a principle.

Bell apparently hoped & expected that if his theorem could ever be applied experimentally to sub-atomic particles, it would demonstrate that traditional views of existence had been vindicated.

It took several decades before Alain Aspect[31] performed a sophisticated experiment in 1981/2 using the *polarisation* of light particles, (rather than simple spin), on the principles of Bell's equations[31]. To continue our analogy and to cut a long story short, these experiments show a significant bias away from 50:50 and therefore seem to indicate that **Einstein was wrong and that nature can manipulate an outcome without a local 'mechanical' cause when a final outcome is required – which may be when the experiment is observed**.

Traditionalist physicists were stunned... along with many others I suspect! (Remember that these results also *preceded* the findings from the Gisin experiments above by over 15years which is why they caused such a sensation).

Somebody presented with this analogy might assume that it demonstrates consciousness more than awareness, and that nature is deliberately trying to avoid our guesses! However the real results aren't as extreme as those you might expect when playing a 'card shark', and they seem to point to something other than consciousness.

The results are still remarkable because they are skewed in a direction that they shouldn't be able to go, (according to traditional science), the 'saving grace' is that they do closely match the subtly different predictions made by Quantum Mechanics.

To understand this we have to go deeper into the actual experiments while trying to bear in mind where the vulnerabilities and opportunities may lie. Remember – the tests compare *predictions* to what happens in reality, and they must remove the potential to skew the results because of distortions in the way that the experiment is conducted.

As an important start point, by using the polarisation of light (alignment) as a measure, the odds move away from 50:50 and were described to me as being closer to 33:67, because you need to measure alignment in 3 ways.

> In crude terms, using a polarising filter, (like some sunglasses), you can divert photons which are aligned in one of those 3 directions to travel along one route, while all others go in a different direction. Different detectors then measure the other two properties to determine their characteristics and overall alignment.

> Aspect's experiment also used 'entangled photon pairs' by splitting photons from a laser and then sending the different halves in opposite directions, first passing through rotating polarising filters before being measured/detected. Although photons can be generated with a variety of angles and spins it is assumed that these factors don't change en-route, and that the rotating filters would eliminate bias across millions of individual instances.

At high level scientists are interested in the number of times that the detected properties for each twin photon **are the same** not different - a 'coincidence rate'.

Different experiments present results in different ways, but to give you a relative measure of scale let's say that the expected 'coincidence hit rate' was $^2/_3$.

> This ratio is based on complicated maths which essentially looks at the different possible outcomes of match or no-match that could possibly occur, (ie. which of the measured angles was different; whether they were aligned in a positive or negative way; etc.). Due to the use of rotating filters etc., it was basically assumed that all of the match/no-match combinations were equally likely and therefore an analysis of those possible match/no-match combinations would give you a coincidence hit rate based on what could happen under the traditional physics of Newton and Einstein.

The expected coincidence rate was also a bottom limit, (ie. the result could be better than this **but not worse**), because there were some additional circumstances where a coincidence/match was guaranteed, (not one that could 'go either way'). These could **only** make the result move towards 100% instead of 66.7% ($^2/_3$).

As a relative measure, compared to the $^2/3$ figure, the actual results came close to 50:50. Under the rules of Einstein and Newton this simply shouldn't be possible. Yet across many forms of the test, the same results have been repeated many times.

As mentioned before, the findings are not as extreme as you might expect if you'd been playing a card shark, (where every expected result would be deliberately defeated). If the actual results were indeed the outcome from a conscious intervention, that intelligent being would be applying its deviousness in a very subtle and disciplined way! We might also expect the results to vary by a significant amount with each experiment, but the results appear to be consistent per a fixed mechanism.

The card shark analogy is also broken because we are not making predictions for each individual run of the experiment – they are being made on the series as a whole.

The other interpretation is that the results closely match the predictions of Quantum Mechanics, which take one other factor into account – that the twin photons may be communicating faster than the speed of light and thereby influencing each other.

As slight differences in the length of cables and positioning of detectors mean that one photon will be detected before its twin, there's the potential for the angles of the 1st detected particle to be superimposed on its twin before that one is detected, and this changes the profile of the coincidence rate to the one actually being recorded.

As late breaking news, on 30th November 2016 a unique worldwide experiment[43] was conducted in which 100,000 people interacted with Bell Test apparatus to randomly change mirrors, polarising filter angles, etc. in conditions where the match results **'could not be greater than 2'** according to traditional physics - but early results show that they were! It specifically tested the influence of human randomness/observation.

From a layman's perspective we can draw the following conclusions: this is more evidence that paired photon 'particles', (and probably paired electron particles)...

- **do** influence each other faster than light.
- **do** retain a non-physical connection which may be likened to crude awareness of each other – but that is the extent of the awareness.
- **may** only determine their properties when we look at them, but this is not a direct implication from the experiments and is still the least proven concept.

The theories of Quantum Mechanics describe this effect as 'Entanglement' but the principle has been extended far beyond the confines of paired particles. The extent to which this is valid is open to debate, (but not here). I should also make the point that **Entanglement is a description; it is not an explanation.** We don't know how it happens; only that it does happen, and that the effect can be used in different ways.

In this respect it doesn't seem to change any of the issues that arise from the Double Slit experiments. For instance, if photons **don't** have any shape or form until detected there is still the question of what causes them to adopt the characteristics of one 'state of being' or another – regardless of how that 'choice' is communicated to its pair.

Translating this into broader circumstances, if the way in which we observe/measure things produces a random effect on one paired particle, our actions would influence outcomes based on the other particle in locations that are far away and seemingly unconnected... but as far as the particles are concerned their actions would still based on cause & effect, because they always react in the same manner. However, the source of the randomness would still lie with us.

Partly for this reason I understand that Bell himself regarded the primary results with sadness, although he did also make the point (in a BBC Radio interview[42]), that there was a way to avoid 'faster than light awareness' if the Universe was rigorously deterministic – ie. if we are all part of a bigger **pre-determined** process, which would still make these results inevitable.

It's an interesting philosophical point but because these results **are** such strong evidence we cannot dismiss the findings.

Strategically, let's remember that nature does present us with a stable reality, so rules must be followed somehow, even if nature's actions are **not** observed by human beings. It is also true that human observation can be made by many different people from many different perspectives & distances. If observation was the only factor, how could stability be maintained? More research is needed before we can form an opinion.

Returning to the philosophical range of thinking, one notion from the Idealist camp is that our reality always conforms to a fixed standard **because** our existence it is being observed constantly by so many people who each share the same stable collective view of how things should work. If correct, this would suggest that if humanity was collectively able to believe in a different form of reality we may actually change the nature of existence!

Yet, once again, we must remember that these effects are only seen in paired photons and possibly paired electrons. While this points to a capability it doesn't necessarily extend to every particle in existence. If it does, that hasn't been demonstrated.

Of course, in relation to the question of Origin we would also have to ask what could observe/measure reality just after the Big Bang, causing it to crystallise in one form or another, before physical beings existed. No doubt those with religious beliefs will turn to an omniscient and omnipotent authority – God.

Others may look for more grounded explanations yet to come.

From a personal perspective the findings from the experiments performed by Alain Aspect and Nicholas Gisin were the least expected findings from my own analysis.

The underlying question is where the principles they have revealed may lead?

We will see later that such a capability could unlock the potential for other very significant mechanisms of origin.

7.1.3   Additional Dimensions - A Way To Square the Circle?

Within science there's a lot of talk about additional Dimensions, Parallel Universes, and 'Multiverses' as ways to explain some strange aspects of reality (as above).

The viability of these concepts rests on both the supposed abilities that these structures may possess, and whether these things actually exist. There's a danger that people may dive-in on the headlines without fully considering the practicalities. So as an example, let's take a look at the first of these structures – additional dimensions.

Let's imagine that the Universe only existed in two dimensions, (width and length), and within that plane of existence there was a square sheet. It had an area of space that it occupied, but it had no depth, because there wasn't a third dimension to occupy. You could also travel to it and away from it in any direction within that thin plane. Other sheets may float in other parts of that same plane of existence, effectively 'side by side', in a similar concept to *Multiple* universes.

If suddenly a 3rd Dimension (height/depth) was created, various things might happen.

The object could remain as a thin sheet within that new dimension, but now, other sheets could float directly above or below, possibly even sharing the same length and width co-ordinates, and just differing by height.

Each of those sheets could form a distinct configuration of existence, without ever interacting, (different sorts of physical reality), and as part of this, the square sheet may even be stretched into a cube, etc. Yet these 'objects' still sit next to each other.

How can we therefore imagine a '*Parallel* Universe', which supposedly has to occupy **the same physical space within our 3 dimensions**?

Some people think of it like the sheets of paper floating above or below each other, sharing the same co-ordinates in the original planes of existence, except that a parallel universe would occupy a different position in a 4th dimension. If we draw an analogy to say that Time is the 4th dimension, then we could imagine different universes occupying the same physical space at different times, for instance.

However others suggest that Parallel Universes have to occupy exactly the same space at exactly the same moment in all of our dimensions, but without those 'cubes' or universes directly interacting.

Unless the sheets could pass through each other, it's hard to see how they could occupy the same space at the same moment – especially if any of them had any thickness. If they lay directly on top of each other, touching, like a layered cake, they would have to have the equivalent of different height co-ordinates. So we find some people arguing for existence in the new dimension as having zero depth – allowing a theoretical co-existence... although to me, a zero depth means that it cannot exist.

We must each ask ourselves whether we believe a zero depth to be possible? If not, then the only way for co-existence to be achieved in all dimensions would be if these sheets were to all pass through each other like fields.

M-Theory[22] suggests that there are up to 10 physical dimensions plus Time, so there would already be 7 hidden dimensions of existence that we do not perceive. Religious believers can therefore argue that the spiritual worlds they believe-in could exist in one or more of those dimensions. Yet, other than this, people do not know what additional dimensions may actually represent.

For the moment, let's stay with the basic scenario and suppose that twin particles sit next to each other on the original 2D sheet but that one of them suddenly travels at the speed of light into the new 3rd dimension. While there would soon be a huge distance between them in that new dimension, the co-ordinates in the original dimensions would not have changed, and for this reason some people have suggested that the particles would, in reality, stay next to each other in the original dimensions – allowing instant communications between the particles. Hopefully you see why I am sceptical about this. I believe the distance between them would be real.

The Gisin experiments have shown that particles can communicate and exercise control over each other **instantly** over large distances, so some mechanism must exist to enable this. Yet is the suggestion of an additional dimension sufficient to explain this?

Some may argue that there could be a physical connection (or thread) between a moving object in our 3 dimensions and its associated data in a hidden dimension.

Perhaps... but I can't help feel that this is in idea too far because of the tangled mess of threads that would arise as things moved about; the fact that we can't detect them; and the likelihood that those threads would be very long, which would introduce a time delay in communications 'down the wire'.

A much more efficient notion would be to extend Thomas Campbell's[28] scenario, and suggest that everything in existence taps into the data dimension like connecting to the web using wi-fi, with an additional super-fast look-up, retrieval & storage mechanism.

Now you may say that this is layering speculation on speculation, and that would be true, but the bottom line is that the experimental results are very real and no other logical explanation has been proposed.

To complete our thinking about dimensions let's return to the nature of reality rather than mechanisms of control, and assume that the introduction of a 3rd Dimension had stretched the sheet into that new space to form a cube. An area of space would become a volume of space, but the nature of the sheet (the material within the Universe) would have been *fundamentally changed by the insertion of that new parameter*, and that might result in new characteristics.

So now let's suppose that a channel opens up between a sheet and a cube, allowing material to transfer between them. If they had both been operating in the same plane of existence then there might be considerable similarities between the material within the sheet and the cube – meaning that the state of any transferred material would be largely preserved. However if the dimensions were fundamentally different they would be likely to make the materials incompatible with the other zone. Any transfer between those zones might cause the material to be de-constructed and returned to its most basic underlying elements as it entered the recipient sheet/cube – possibly returning to a state of pure energy. **As a lead-in to later chapters, we may see here the beginnings of an alternate explanation for the Big Bang**.

In overview, additional dimensions provide the theory of origin with 3 things:
  i)   a place for material to exist outside our 3 dimensions,
  ii)   the opportunity to have other mechanisms of origin, and
  iii)   a way to generate effects such as Universe Inflation.

In terms of more current issues, we can also see that additional dimensions may help to resolve the EPR paradox and the nature of paired particles, but we shouldn't get over-excited and lose sight of the practicalities of what these suggestions might mean.

7.2 **Philosophical Implications of the 'Double Slit' and EPR Results**

Through the course of Chapters 6 & 7 we have found that ideas being proposed by scientists to explain different findings appear to be blurring the philosophical lines between Materialism, Idealism and Dualism.

The main findings/implications which challenge the Materialist perspective are:

- any requirement to *interpret* circumstances rather than simply reacting to a physical influence, (eg. determining the certainty of the 'which path' information, or whether there is a need to crystallise into a particle).

- suggestions that photons & electrons are able to reach out and sense their twins, or their environment, (eg. whether 1 or 2 slots are open).

- suggestions that messenger particles are influenced by an exchange of data/information rather than any physical contact or interaction,

- any suggestion that 'signal' photons predict the fate of 'idler' photons.

This is because such factors either: deny causality; or break some of the fundamental Laws of Physics; or require properties which Matter/Energy shouldn't have.

Having said this, Thought is real and if it has to be explained entirely within physical Matter/Energy, the capabilities of entangled paired particles may point the way to do so. Yet this might also open the door to other explanations based in other stuff.

While science will ultimately follow the evidence, there is no doubt that there is a raging argument over how that evidence is interpreted by advocates of the different philosophies. The danger of misinformation is compounded by exaggerated claims.

Mathematical models (either in support or denial of a principle) are useful but they are not proof. They require verification against real world evidence. Equally, properties exhibited by paired particles across vast distances is something that can be latched-onto by Theists, who believe that God can receive prayers and monitor/influence all activity in creation without being present in all places. But can we honestly say that all physical existence is made of entangled paired photons - one with us and the other held centrally by God? Even if this points to a general capability, it doesn't prove God. The best it would do is re-open the potential for a capability long denied. This is partly why such communication is so heavily denied by some – generally using maths.

So in combination, we have to ask ourselves whether the Double Slit and Bell Test experiments are the smoking guns which could unlock the mystery of existence.

The fact that these experiments are repeatable, implies is that nature is still following rules. Equally, the increasing prominence of data and information within theories of existence could help to demonstrate how the stability of Matter/Energy and the spontaneity/creativity of Thought might work together.

I am not alone in being wary of new theories which tell me that I should ditch the experiences of a lifetime in order to resolve some obscure fact, but there is some comfort in knowing that a number of these real world discoveries have common themes.

All views say that we live in an illusion. It will be interesting to see if this marks the point where religion and science can begin to find common ground, or they continue to maintain their separation.

8   If Creation, What Was Created?

In earlier chapters we have seen the outline logic which has led scientists to believe that our Universe began 13.77 billion years ago. There seems to have been an event at that time, but the nature of that event is still open to question because a number of recent findings have undermined traditional thinking about what may have occurred.

Various ideas can help us to refine our opinion.

Firstly, we should recognise that the term 'creation' can be applied loosely - to simply mean a beginning, but we have also seen that this could also be interpreted in two ways: as a fresh start from nothing, (pure creation), or a dramatic next step in an inevitable process (transformation). We don't know which of these the Big Bang may represent, but it could in fact be both, (pure creation alongside something that exists).

Secondly, if we identify the need for new capabilities, (beyond what is known to be achievable), we have to explain how those capabilities arose. A failure to explain those capabilities as part of an inevitable process must lead to us look for spontaneous, random, or external influences.

Thirdly, we need confidence that we have a reasonable way in which to look back into the past, and the basis for doing so comes from 2 perceptions :

- that our *physical* environment only seems to evolve on the basis of rigid rules which use the principle of cause and effect – implying that events can only unfold in one way,   and

- that our current reality is a precise end point and, in order to be viable, all explanations of origin must bring us to this exact state while accommodating **all** known factors over the past 13.77 billion years.

If we adopt a reverse engineering approach on this basis, then we might be able to envisage the state of existence immediately after the Big Bang event – but **only** if we are confident about two things relating to the intervening period:-

i.   that the capabilities of Matter/Energy haven't changed.

ii.   that we are aware of all significant factors that were involved in events over the past 13.77 billion years.

In considering point 'i' astronomers observe a Universe that seems to operate according to just one set of rules, but physicists suggest that physical matter could have achieved stability in many other ways, (with different configuration settings). The melee of an unregulated explosion might produce many of those different forms of existence alongside each other – possibly one per isolated galaxy. So within the theoretical range of possibilities, there may have been no change (which would be good news), or a variety of changes (which would be very bad news for our reverse engineering), but we need to end up with the single uniform existence we have today.

In terms of assessing our ability to look back into the past we need to focus on events after the Big Bang which also occurred on a Universal scale; and there is only evidence for one such thing – the 'sudden' acceleration in expansion. Yet this isn't normally attributed to a configuration change, it is thought to relate to large physical influences *within the normal rules of existence*, (the Laws of Physics as we know them). Our ability to look backwards seems to be preserved.

Unfortunately, point 'ii' above suffers from more immediate concerns, because we don't know everything about the *current* state of existence, let alone the past. The reliability of any backwards perspective is therefore challenged.

Some people may therefore conclude that it is pointless to try and look back to the earliest moments of our Universe as any speculation could lead to a near-infinite set of possibilities instead of the single reality we imagine from causality. Yet that isn't entirely true because it depends on the level of accuracy that we hope to achieve. If we stick to broad principles we may still achieve some worthwhile insights... and besides, we are creatures with curiosity.

Even so, we need cross-checks and markers to give us a level of confidence that our theorising is on the right track, and we get these yardsticks from various sources.

We have various rules, (the Laws of Physics), which determine how we can speculate about a prior cause. We are also aware of the types of mechanism that nature appears to be capable of. Yet in choosing which might apply at any stage, we must formulate an opinion about what has to be achieved.

Overall we aim to build a picture of the features/characteristics which might exist immediately after the Big Bang. A number of basic questions help us to do this, contributing to our backwards view, and not surprisingly these all point to some of the key factors that may shape our reality:-

a) What is *the nature of Time*, and does it impose any additional requirements?

b) Is anything required to enforce the *uniformity* that we can see across the entire Universe, (a single set of rules for existence), rather than having different types of reality? As part of this, what form do the *settings of reality* take, and how did they arise?

c) Does physical existence require a **structure** in which to exist, such as additional dimensions or even a 'fabric of space'?

d) Chemical reactions seem to occur naturally because of the way that atoms are structured and yet this doesn't explain how *repeatable processes* involving a complex series of steps can be undertaken regularly by unthinking chemicals. Is a template of some sort required?

e) We see examples in nature where unthinking things appear to adapt themselves to different circumstances. What allows dynamic *control* to be exercised?

f) Events normally require *triggers* which may take various forms, so what was required for the Big Bang if it was not spontaneous pure creation?

If 'a-e' have to emerge from the environment immediately after the Big Bang then we can also begin to speculate what could put them in place, and also how they might be deployed later. 'f' is trickier and can only be speculated about in the most general way.

However there are other strategic factors which are closely linked, most notably :

- explaining the size of the Universe today in the timescales available.

- the starting conditions for the Big Bang, (state of the prior Universe).

- the need to avoid luck in achieving our particular flavour of reality.

We have already seen that the mainstream suggestion, (about a highly compressed Universe prior to the Big Bang), has a number of problems with it – not least of which is the fact that it can't achieve our size of Universe without breaking the Laws of

Physics. However, that isn't the only option. What if it was originally spread out quite widely instead, and there was a different type of Big Bang mechanism?

There are several ideas along these lines which overcome some basic problems without resorting to suggestions like Cosmic Inflation. They all deserve a fair hearing to see how well they fit with the other characteristics we can identify.

Amongst those features we need to resolve a dilemma. The primary way in which people have sought to explain how our version of reality was achieved without luck is to argue that nature has a way to experiment with different configurations – yet we see that the Laws of Physics force Matter/Energy to *conform* to the rules of existence and not to rewrite them. Special conditions must be required to achieve such changes if they are necessary to make a theory work.

Our logic has also suggested that a small number of **other** types of stuff may also underpin existence, and if that is correct these must work alongside Matter/Energy to shape our current reality. One in particular stood out, based on the opposite but complementary principles of *spontaneity* (without cause) and *randomness* (more than one possible outcome for a single start point).

Overall, such capabilities must help us to overcome the 6 impossibilities which seem to dominate the subject of prime origin, (per Chapter 3).

You can hopefully see the factors which shape opinion about the outputs from creation.

When considering the nature of change in the Universe there only seem to be three generic types of activity:
- Pure Creation;
- Complete Destruction, and
- Transformation.

In this sense, almost by definition, any act of 'Pure Creation' must ***add something*** that didn't exist before to a starting position, while 'Complete Destruction' would involve something being ***entirely removed from the pot of existence***.

To our knowledge mankind has <u>never</u> witnessed either of these events; it has only seen Transformation from a structured form, (eg. matter), into a less structured form (eg. energy). According to science, the same amount of underlying stuff remains in existence before and after any reaction – including nuclear explosions. Nothing fully disappears. 'Transformation' leaves the size of the pot unchanged.

Having said this, there is a rationale within the Big Bang concept that may point towards a rare creation event... but there is currently no real driver to search for things that may have disappeared completely, (there's nothing to look for).

In order to believe that Pure Creation may have occurred we need to look for things which have been added, and that isn't possible if it is argued that the Big Bang marked the starting point for everything. We would merely have to believe in the concept without any prospect of getting conclusive evidence for a prior absence of anything.

Yet we have seen that the latest scientific findings point to a sudden injection of Dark Energy 8.77 billion years after the Big Bang event that may have to come from somewhere outside the original material within our space. While some will speculate there could be transfers of energy from other parts of existence, which is unproven, pure creation would seem equally likely, except that it is far more difficult to rationalise.

If pure creation is possible at all, why shouldn't it happen many times? If there is a limiting factor (possibly special pre-cursor conditions), we should try to identify it. Otherwise the absence of examples may be a good rationale for believing that it doesn't happen. However there are lots of potential examples, it's just that the prevalent philosophy of today labels them in other ways. So which view is correct? We need to consider some specifics – for instance:

- As we saw in chapter 4 – do all supernovae represent exploding stars, or could some of these events be pure creation?

- New thoughts that pop into our heads could be a very common example, marking a truly new beginning that adds something to the 'pot of existence', and may also turn existing raw materials into a new machine.

I find it interesting that social convention generally applies the word 'Creation' to physical Matter/Energy and rarely to Thought, and yet thoughts must come from somewhere. I don't believe that anyone suggests that we each have a dormant stock of blank thoughts which can be turned into useful 'populated' thoughts.

> So where do new thoughts come from? Does this represent genuine pure creation or just a use of electricity? Electricity on its own doesn't have ideas, which is why many people feel it is simply used as a physical 'vehicle' which translates the effects of 'another type of underlying stuff' into physical actions.

Returning to the specifics of the Big Bang, if we take the 'blank canvass' approach, the number of things which may have to be created could either be very large or quite small, depending on your philosophy or beliefs. To illustrate the range, we can tap into the 'traditional' ideas associated with Religion, Philosophy and Science. Taking the religious angle first :-

- People who consider that God designed every individual thing throughout time, and directly brought each one into existence, are saying that God the Creator had a very large list indeed.

- Others speculate that if God was the most intelligent being ever, then we might suppose that He was also clever enough to design and implement the minimum required that would allow existence to evolve and generate all of the remaining detail itself. For instance, the philosophy of Deism adopts this line of thinking - implying that God would have created the mechanism of Evolution.

  Science is more inclined to this view of increasing sophistication even if it doesn't advocate God.

However within this we find that some religious texts suggest that God initiated pure creation, while others suggest that He transformed what was there before.

Science as a whole is open to both possibilities and doesn't take a position on this issue, but it does emphasise that the raw underlying stuff of existence naturally evolved into the complex Universe we see today. For science and atheism, if there was a true beginning, then the outputs of creation were probably very limited in number.

When trying to assemble a list of candidate factors that might have been outputs from the Big Bang, Chapter 4 identified 4 fundamental 'elements' which might underpin our reality, and while others may emerge, for the moment our attention focuses on:- Energy, Force, Thought, and Time.

In addition to this, I listed six additional factors, ('a-f' above), which may be necessary to shape the substance of existence. If correct, these factors may have also needed to be created. We need to look at all of them, in order to form an opinion.

8.1.1   Time

The future isn't what it used to be !   (Anonymous)

People believe in Time because of what it does in two major respects:

a)   The forward progress of Time gives us sequence; a '*before*' followed by an '*after*', without which there could be no movement or Thought.

b)   The pace at which Time progresses also gives us our yardstick on how fast things happen, and indeed the rate at which we age, showing us how long we might live – at least in physical terms.

As already mentioned, Time is perhaps a bigger mystery than most other factors because in relative terms, it's harder to imagine what it may be. There seem to be 3 main possibilities :

- Time may just be a concept, so it ***wouldn't*** control movement or the rate of change, it would only be a measure of it – ie. a human construct,   or

- Time is like a force which can only move forwards to influence events but not to store them. The core philosophy here is that '**there is only now**' and people can regard the influence of Time as fluid and without structure. Alternately you may think of Time as an additional dimension within existence, (like a slope we can only go 'down' - providing a type of force).

- Time may be something more tangible, having a structure or mechanism that is populated by 'events' which might be re-visited.

   Some people have envisaged a physical snapshot of existence being captured somewhere to effectively allow you to land on something solid in the past if time travel were possible. Others have conceived that if nature configures itself in accordance with data that is held about it, then the mechanism of Time could be like a video tape with data/instructions on how to configure existence at any moment in the past or future.

   Either way, that structure might be linear, (drawn as a ***time-line*** that extend forwards and backwards without end), **or** it might be circular, in the form of a loop or ball – per Chapter 5.

To many people in the West, and generally to followers of the Judeo-Christian tradition who believe that Time is linear, it can seem strange to imagine that Time is shaped in a loop or ball. Yet the cyclical nature of events is a very old concept stretching back to the Babylonians, early Hindus, and Buddhists as well as being familiar to other cultures such as the Mayans and Incas in central and south America. However a cyclical process, (like birth, life, death and rebirth), doesn't necessarily mean that events will exactly replicate themselves on the next round; but a true loop of Time **would** imply this.

Linear and circular concepts will both seem natural to those who are brought up to believe in them – ie. that their perception is the truth. Interestingly, few religions seem to argue that there was a start to Time although this was part of ancient Greek thinking, (ie. 1st movement). Yet spontaneous pure creation may represent such a start.

If there was no movement or sequence **anywhere**, including no movement within atoms, Time could be said to stop. Indeed, many scientists and philosophers use this logic to say that the Big Bang was the start of both time and physical existence... (however that partly depends on whether there was any sort of existence prior to the Big Bang). Their underlying point is that if Time cannot be measured, it either cannot exist, or it becomes meaningless. For example :-

> If everything did become entirely static, (including atoms and light), and was then re-started after a while, we would never know how long the 'gap' had been unless we could measure it – even if that was by reference to God twiddling His thumbs for a while.... which would still imply that there was movement somewhere, and Time had not stopped entirely.

> Put another way, if time stopped completely and everything was totally still, including God, then the gap would truly have no meaning because nothing would change, so there could be no reference point to identify the gap and nothing to attribute significance to it. Yet this also begs the question of what could re-start it – a spontaneous influence, or could Time be on a timer?

Yet others argue that Time can never stop. An 'Absolute' form of Time would always continue to 'tick' (move forwards) even if all physical movement had ceased, and it would progress at a fixed pace.

Unfortunately nobody has found any reference point for Absolute Time, so any measure of Time that we take is always established as a cross-reference to other things which would themselves stop if Time stopped. In other words, Time is relative, and from this Einstein and others have used equations to show us that the pace at which time progresses might change under different circumstances. We'll consider this later.

Having formed our 3 ideas about what Time might be, we can now contemplate whether it is an independent factor in existence rather than a symptom of some other stuff, and whether there could be a start to Time in any of these circumstances. The answers have direct implications for all other things that underpin existence, eg.;-

- If Time is eternal does this imply that Matter/Energy or Thought would also have to be eternal? No, because their creation, as an event within Time, could occur at any point on the timeline.
- As an alternative, if Time was not eternal and had a beginning, does it imply that Matter/Energy or Thought could be no older than Time? Once again – No it doesn't. Such things might exist before Time began, but they would have to be entirely static until Time/Movement commenced.

From this thinking, let's consider what would be needed in order to start Time. Firstly, if Time was a symptom of either Matter/Energy or Thought, not something in its own right, its start point would be marked by the first movement in either one of these. Yet there's a twist here, because if both of these factors were entirely static there could be no earlier dynamic force to provide the source of movement that would kick-start Time. Once reactions began they could escalate and develop their own momentum, but a *start* to Time would **have to be the result of spontaneity** - a type of pure creation.

This is equally true if Time were something distinct and an enabler of movement. Nothing could trigger it, because all triggers need movement. The only alternative is that Time would have no beginning, (ie. be eternal).

So lastly, if we perceiving Time as a series of moments which are stored and can be visited. the notion of a 'Ball of Time' suggests that there is no start point to a circle, but would imply that the structure of the ball may need to be put in place somehow, as well as providing the first movement.

Logic suggests 3 possibilities :-

i) the structure of Time, and the events which it contains have spontaneously appeared – with or without separate beings who choose their path around it, or

ii) the structure of Time is eternal but the events within it could emerge spontaneously to inflate it from small beginnings.

iii) both the structure of Time and the infinite options within it are eternal.

Option 'ii' effectively suggests that the smallest circle or ball is a dot – a structure of sorts, but with nothing in it. A means would then be needed to inflate that dot into hoop or ball.

Although an inflation of Time seems to fit with scientific ideas about the inflation of the universe, such notions still require something, (eg. the dot), to be in place without beginning or to emerge spontaneously. Even if it had existed forever, if the dot only began to inflate part-way through its existence, that change to its previously eternal status might require spontaneous pure creation.

The significance of this is that if spontaneity comes from a different type of stuff then Time would now be subject to other influences, and not just be an influencer itself.

In short, any start to Time has to be the result of spontaneity/pure creation to change an eternal static state, but there is still the possibility that dynamic Time could be eternal.

At this stage I'd like to point out two other characteristics of Time before we move on:

- No concept of Time seems to suggest that Time is able to **create** either Thought or physical Matter/Energy, so *their* origin must be explained by other means.

- Time only seems able to move forwards. There's no conceptual way, within our scientific knowledge, to change the past.

8.1.2   Settings for Reality

Once the underlying stuff of physical material exists, scientists believe there are only a small number of basic parameters which fashion the style and characteristics of the universe to determine how things operate and interact within it – possibly as few as 6. However each of these can theoretically have a myriad of settings. For instance, the number of dimensions which might exist within reality could be just the 3 that we know, or potentially number in their thousands or billions.

If these 6 factors represent the core configuration of reality, (effectively the 'dials which God could turn on the control panel of existence', to shape our environment and ourselves), then we should consider whether these settings were themselves inevitable, or had to be created.

To illustrate what we mean by 'settings for the basic parameters/laws of nature', we can see that forces not only exist but they operate at consistent levels of strength. For example, in terms of the setting for gravity, **something the size of an orange could have been given the same power as we currently see in a planet.**

Theoretically, in the earliest moments of the Big Bang when the Laws of Physics were still being formed there seems to be no apparent reason why the strength of gravity shouldn't have been set at a much stronger or weaker level, or even able to change its strength under different circumstances – perhaps periodically, say on Tuesdays (☺) .... or following some other cycle.

However in reality there's no astronomical evidence to suggest that gravity has a different strength in other parts of the universe, as the motion of all stars/galaxies appears to conform to the same equations that we apply here. Everything seems to work to the same settings – and although that may just reflect the assumptions being made by astronomers, it's the best we have.

So out of the melee of an unstructured explosion we have to ask : i) what enforced that uniformity, and ii) where those 'settings' in the fabric of physical matter came from?

> Incidentally it's a very good job that gravity is pitched at the level it is, because if it was even slightly stronger it's believed that suns would burn out before there was enough time for life to develop on the planets around them, and if it was weaker, it would have taken a lot longer for material to have gathered together to create the suns which would also have burned a lot less brightly, or may not have burned at all – preventing the formation of the heavier elements and our solid environment.

> The strengths of other forces also seem finely tuned to allow atoms to forge a variety of new compounds, instead of forming bonds that were too strong to allow further reactions, or too weak to react at all.

The universe appears finely tuned as we seem to exist in a very narrow range of those possible settings that would permit our reality to occur in the period since the Big Bang. This narrow range is sometimes referred to as the 'Goldilocks Zone', however we believe that many other configurations might produce a stable reality even though they wouldn't resemble our Universe.

So what are these 6 parameters on the control panel of existence? In short they are :-

- the 4 Forces
- the number of dimensions in physical reality, (considered to be one factor).
- the density of the material in the early universe, moments after the Big Bang, which had to be sufficiently uniform to allow the shape and evolution of the universe to occur.

A seventh has also been suggested : the inherent speed of Time, or the rate at which things interact or become older.

Yet from another perspective it could be argued that there may only be **one way** in which those different factors might combine to produce a stable existence. In other words, they may only be able to co-exist in a stable form when they are pitched in this one way. This **doesn't** seem to be the general scientific view where many stable options are considered to be possible, but if we run with the notion of a single inevitable reality, within the melee of the Big Bang, different arbitrary setting which may have originally formed could play off against each other, to bring each of the 7 factors into line with that stable balanced position – ie. a self levelling mechanism.

We simply don't know enough, however many people presume that new settings, (ie. new rules of existence), could only arise under the extreme conditions of a singularity where everything is reduced to its most basic and unstructured form of existence.

For our purposes the exact number of parameters isn't really the issue. The underlying point is that there are a **small number** of key factors which seem to have the potential for different settings, but which managed to achieve a perfect combination against very high odds.

8.1.3   A Framework for Existence & the Fabric of Space

For those who are new to the subject, a Framework for Existence may seem like a strange idea, but the concept does help us to draw out some useful points, and scientifically it may yet be shown to have some truth!

I have already introduced you to the competing notions that

- if space can be filled then it is something... ie. physical matter from the Big Bang is occupying a space that exists,   **or**

- if there is no existence whatsoever beyond the outer reaches of the Universe, then space may have to be created before physical matter can fill it.

These ideas effectively suggest that space itself has a fabric, however the idea goes much further than this, to consider how existence regulates itself to bring uniformity, and ultimately this allows us to explore ways in which the Universe might be shaped.

Science perceives that in the first moment of the Big Bang all of the matter and energy which the Universe has ever contained, emerged in its most basic state with absolutely no form, structure, or properties.

Later, during the first microseconds of existence, it is believed that the force of gravity was established at just the right strength to make it work effectively, (although we don't know how), and shortly afterwards the key properties of other forces, energy, and particle matter emerged, which ultimately led to the creation of the physical Universe and ourselves.

As touched-on already, a basic question in science is:

What causes all matter/energy & forces to adopt, and stick rigidly to, a **single set of physical laws** that give it stability, and consistent patterns of activity? Why didn't we get a variety of configurations?

The simple answer is that we don't know. In referring to Quantum Mechanics, (which starts with the premise that all outcomes are possible), Albert Einstein famously said

"I am convinced that God does not play dice"

Einstein and other scientists have proposed that certain outcomes in the real world require hidden guiding influences which operate on a large or even a universal scale. The 'de Broglie–Bohm' theory[33] is a case in point, suggesting that a guiding field or wave may exist across the Universe. In truth this theory hasn't found widespread scientific support, mainly because of its 'non-local' operations, (ie. it requires something to *impose* discipline rather than each tiny piece of existence controlling itself)... and, for some, it also smacks too much of God.

Whatever factor brings about such uniformity, it represents a framework for existence which may have needed to be created, and/or configured. There are various ideas about what a Framework for Existence might be, and these follow one or other of the 2 main themes above, which can be re-stated as :-

- an **'external'** structure in which physical matter/energy can be contained, and through which standard shapes and patterns of activity can be <u>imposed</u>, which might also mark the boundaries of the Universe and therefore of existence, or
- an **'internal'** mechanism that allows each sub-atomic particle of matter/energy to <u>regulate itself</u>, (ie. to impose the laws of physics on itself), using the principle behind messenger particles which are already suspected of instructing objects how to configure themselves.

Both are effectively ways in which the settings for existence (last section) can be applied/imposed in a consistent manner across the Universe. We would otherwise have to rely on luck to achieve uniformity, or we would have to say that there could be only one shape which existence could adopt in order to achieve a natural balance.

Yet to guarantee uniformity *across the entire Universe* there is a presumption that:

- the same mix of elements was created everywhere, (ie. you wouldn't get uniformity at the touch points between different zones, (eg. if our visible Universe was one of those zones)), **and**
- these materials can *only* become stable in one configuration – either because of a self-balancing mechanism or because there is only one stable possibility across all configurable settings – neither of which are not proven.

A more specific concern for the 'self-balancing' theory is that it's hard to see how some of the factors could flex/adapt; for instance how could you get rid of a physical dimension once it existed? Perhaps it's sufficient to only flex some of the 7 settings?

The alternative is to consider that many other permutations of existence are possible, each governed by different strength settings on the control panel - in which case we would again have to ask

- If different parts of the Universe initially adopted different combinations of settings, how did nature allow different flavours of reality to co-exist; or prevent itself from being in perpetual turmoil; or prevent the Big Bang from fizzling out?
- How did our part of the Universe, (and potentially the whole Universe), manage to maintain the configuration it ended up with – rather than changing again?

These questions remain unanswered, but once you understand them they become a useful test of any theory about the early universe.

For some, the answer lies in the possibility that other forms of existence do operate in different galaxies and that the emptiness of space isolates the different effects. Others may turn to the possibility of an internal or an external framework. So with these points in mind, let's move forward by considering ideas about an **ex**ternal *physical* framework which might impose structure.

If physical material floats in space, could 'space' itself be the framework? We can think of space as being **infinite** and that physical material is floating outwards from the point where the Big Bang explosion occurred, to occupy that space.

An infinite framework **cannot** be achieved by an explosion. An explosion is always a **finite** size – even if that's very large. Therefore an infinite framework would either be eternal, (no beginning), or would have to be put in place by something that was already infinite – God? Yet there's an alternate view.

Both space and the Framework could have a limited size. As we have already seen, some scientists have suggested that the Universe is a **finite** size, (only existing up to the 'boundary of the ball'), and that 'space-time' is some sort of structure that's being inflated. On this basis the framework could have potentially started-out very small and was inflated along with the rest of the Universe. Alternatively, it may have already been large - either as a leftover from a previous Universe, or because it was created that way by something else....

Hopefully this illustrates the range of thinking.

So is there any further evidence from science to point us towards an external framework as opposed to an internal version? At headline level I feel that the evidence either suggests that **both** may be active factors (ie. neither can explain everything – we need elements of each), **or** that it now favours the **internal** option, but to understand why we have to delve a bit deeper.

One of the great mysteries in science is the nature of Gravity. Unlike the other forces which seem to use the same type of messenger components to transmit their effects, no equivalent particle has been found for Gravity, (ie. no 'Graviton' messenger particle has ever been discovered, and many believe that it may never be detected). So we don't know how Gravity works.

Without particles, one of the ideas to explain how it might operate has been to suggest that it may represent a distortion in the fabric of space-time, (ie. a distortion in the external Framework of Existence).

> Imagine that the empty space in the Universe is like a huge sheet of elastic. Then imagine putting heavy metal balls on that sheet. At those places where the objects sit you will see the elastic stretch and sag, producing a dip in which each ball rests.

> Then imagine that other much smaller balls are trying to roll past the larger ones. As they reach the dip in the elastic they begin to roll towards the larger object, being pulled round and round into the dip. It is that pull which science equates to gravity.

I love this analogy. It's such a vivid and visual impression of what might be happening, and I have even seen it demonstrated. It's great... however it doesn't explain reality.

Firstly, even if we accept the basic premise that space-time, (ie. the elastic), represents a physical framework that can be distorted in order to provide a place for things to sit, we have to remember that in the emptiness of space things appear to 'float' so there would be no downward pull to cause a dip. Without an inbound force we would effectively be saying that gravity is the explanation of gravity. It's not a real explanation.

Secondly, the elastic wouldn't be a sheet – it would be everywhere, so any distortion would exist in all directions (not just downwards).

Furthermore because the object was forcing the elastic material to bend outwards in order to surround it, the associated forces being implied would be outwards not inwards – a force that repels not attracts?

Yet the notion of a distortion in the 'fabric of space' remains a useful way to model the effects of gravity, and in particular for our purposes it's the potential to 'curve space' that is the key idea here. If it were true, it might be further evidence for an external *physical* Framework of Existence.

If, for a moment, we follow-through on the idea above, that a curved fabric of space might be a force that repels, then it might be able to explain something else – nature's tendency to fill a void.

We saw in chapter 4 that science doesn't accept that there is such a thing as a 'suction force'. Pressure differences will typically represent a difference in energy levels where the flow will always be from high pressure areas to low pressure areas. We rationalise this by saying that nature seeks to remove an imbalance, but is this the only factor? Perhaps distortions in the fabric of space can add another aspect to the outward force of the original explosion to maintain or even accelerate those effects, (if this effect repels not attracts)?

In the analogy of the elastic sheet going round large objects this would be the equivalent of saying that if formed a hill instead of a dent, (seeing the elastic from the other side), or that a large spread-out mass would compress the fabric of space to cause a large 'density slope' which could cause things to move outwards. Could this be another way to explain Dark Energy, or even part of the driving force for the Big Bang itself?

Following a different line of thinking, the 'fabric of space' idea has been used as a way to rationalise why the Universe may be finite instead of infinite, because it might be curved into the shape of a ball.

If the notion of a curved fabric of space was correct, it might prevent you from ever reaching its edges, or even seeing them. For instance, if you travelled very fast in what you thought was a single direction, you might never reach the edge of the Universe because the fabric of space would always bend you into a circular path towards other parts of physical existence... or so it is suggested.

> Put another way, if it were correct, then when considering the entire universe, the angles of a triangle wouldn't actually add up to $180°$

It's a nice theory, but would have been extremely bad news for astronomers as so many of their calculations, (which are based on triangles), would be wrong. However recent observations from the WMAP satellite provided evidence that space is not curved, (which is where we finally come to the example of scientific jargon which I presented in the first pages of the book).

> *The 9 year results of the WMAP survey concluded that the universe is flat with only a 0.4% margin of error and that Euclidean geometry probably applies.*

While some cynics suggest that this is 'astronomy tying to protect astronomers' – it's the best that we mortals have been offered.

The undeniable point is that reality shows that if there is any curvature it will be much less than many expected, if it exists at all. I suspect that the majority of readers, (like NASA), will believe that space is flat and therefore potentially infinite, **which could make an 'external' Framework for Existence less likely**.

We should now consider ideas about '**internal**' frameworks.

The easiest way to imagine an internal framework is to say that every sub-atomic particle contains a self-configuring mechanism which would allow it to change its properties if any new settings/instructions were communicated to it.

We already know the theory that most forces communicate their effects through messenger particles, therefore any objects which receive those messenger 'instructions' would configure themselves differently. (The basis of a framework?)

This is similar to the concept of Object-Oriented design in computing.

In this programming technique, encoded ideas known as 'objects' move around a computer system in discreet self-contained packets of data that can allow each object/concept to fully describe itself, enabling its image to be drawn, or its properties to be used wherever that object ends up in the system. In other words, if anything encounters an object it will say 'Hi – I can do this'. As an example, if the concept we were describing was a cube, that bit of code floating around the computer might :-

> ... describe the structure of that 'object' by saying it had 8 corners across 3 dimensions; that the angles of those corners were all 90°; and the length of each edge would determine its size.

> By inserting new values into some of the parameters which the object carried around, (perhaps using 'messengers'), we could perhaps lengthen some edges to make the object more rectangular than square, or even distort the shape by using angles that were not 90°.

> Wherever it went within the computer, the object would be defined by the settings/descriptions that it carried.

If physical particles acted in the same way then each particle would configure itself using the settings which it carried. On this basis, **large scale transformation of the physical environment** could be achieved by issuing a general instruction to every particle in existence to adopt new settings, and then waiting for each particle to conform to the new standard.

The evidence for Messenger Particles; the effects recorded in the Double Slit experiments; plus the results of the Gisin[19] experiments in Geneva, (which show that instant communications across the entire universe is a possibility); all add a degree of validity to this concept.

Could this represent an alternate explanation for the Big Bang?

While such a mechanism would represent an 'internal framework for existence', we would still have to determine where those new instructions/settings came from!

8.1.4   Processes and Control

Processes are a series of specific events that have to occur in a particular sequence, and they generally have to be performed on a regular basis. Even if the right 'ingredients' were available for a process, why should the individual steps be consistently processed in one way rather than another? If events were to take place in a different sequence they would represent a different process that almost certainly resulted in a different outcome, (eg. a different gene sequence).

So what causes a very specific and complex sequence to be followed/repeated? This is particularly relevant for very long sequences, such as the construction of proteins.

In answer, people have speculated that either :-

- the necessary elements or constituent parts are somehow constructed in a way that restricts them to one particular sequence, (only one way is possible), or

- there must be a controlling influence/template that makes them 'follow the path'.

Aside from this it would just be a matter of luck that things unfolded in the same way, and luck is not a process. Luck would also mean that if something complex happened once, it may take a very long time before it occurred again... if ever.

Science argues that all things must begin in basic ways and then have the time to evolve into a more sophisticated form - yet that gives atheism a big problem. Scientists[34] believe that 13.77 billion years simply wouldn't be long enough for our Universe to emerge from random events, so materialists puts their entire faith in natural **_processes_** to square the equation, (without a guiding hand). However we then have to explain how **_repeatable_** processes can emerge naturally, without design.

Natural processes have been well studied in our environment, and they don't always require Thought to put them in place. We know that many of these processes existed before any creature that is known to have lived on Earth, so we assume that they came about before the influence of any physical creature anywhere. Such processes include the movement of tides; the formation of crystals; and the photosynthesis of cells (which had to exist before living plants existed); etc. Scientific research has shown that these chemical processes happen for very specific reasons.

Yet there are also many processes which we observe and yet cannot explain, from the dual slit experiments to many of the complex processes of life. Here in particular we find processes which are *assumed* by science to have arisen inevitably from the actions of physical Matter/Energy but which are **not known to any degree at all – not even in concept** – such as the formation of the first cell. That's a very big gap. This is also explored in Book 2 of this series where I consider Life, Mind, & Soul.

Whether you consider the Big Bang to represent pure creation or a fundamental re-shaping of material which had been there forever, *'processes' have to emerge after the initial moments of the event*, based on the configuration of existence which emerged.

It is believed that *no mechanism or pre-determined sequence could have survived the transition of an all-encompassing Big Bang explosion*, even if there had been a prior existence. So if some processes are not based on chemical structures, (because those didn't exist immediately after the Big Bang), what else could have set the template for a complex series of steps? Other than Thought & design we do not know.

Yet anyone interested in the subject will inevitably want to consider the possibilities, so in looking at how processes may emerge, we have to distinguish between :-

- those factors which cause physical matter to operate in particular ways, and
- those remarkable coincidences where many disparate things seemed to have come together at the right moment.

We've already considered basic notions about what could set the rules of physical existence, so let's turn our attention to examples of apparently remarkable co-incidences to see if luck is the only way to explain them. For instance, do physical processes explain how exactly the right circumstances for life came to exist on Earth?

We live on a planet that is endowed with a relatively huge amount of water, which is also just the right distance away from the sun for the water to remain liquid, (not frozen or turned to steam and leaking into space).

In addition, the planet is rich in carbon which is just about the only element that has the right chemical bonds to enable the necessary reactions for life, and it all exists within an atmosphere that is conducive to life.

The sun is stable as well as being the right type to nurture life instead of frying it, and our path around it is nearly circular, (not oval as is often seen in other worlds), to give us a modest range of regular temperatures which can support a variety of living creatures and plants. Our planet spins at an angle that always tilts one way which serves to distribute the heat from the sun across more of its surface. It also has another very rare characteristic in that it is shielded from harmful solar radiation because it has an iron core that generates a protective magnetic field around us.

Once again you can see why people believe that we live in a narrow 'Goldilocks Zone' where the conditions are just right. However, other remarkable factors exist about our circumstances even if they aren't related to survival.

We have a single moon that regulates the tides and influences our weather systems. It is also, (in a remarkable quirk of fate), exactly the right size and distance away from Earth to just ***perfectly*** cover the sun during an eclipse, (which first allowed us to study the sun's corona and gain our initial insights into the workings of our nearest star).

I could go on, but for the time being let's just accept that there's rather a lot of beneficial factors and their existence has to be explained without resorting to luck. The standard answer from science has two elements, ie. that:

1. the Universe is so vast that every outcome (ie. each possible mix of the factors in existence) could inevitably arise somewhere, (within the laws of physics),

2. where circumstances are right, naturally occurring processes will make some complex things inevitable.

In short, if life was possible at all, it would inevitably arise somewhere in the universe because it is so vast that it can accommodate every permutation of circumstance at some point in time. Life happens to exist here because the Earth is one of the few places where the necessary combination of circumstances was right.

The difficulty with this type of logic is that it is fine for a one-off event, but not where a complex series of steps has to be repeated exactly and on a regular basis.

There is still a need to explain how complex processes could arise naturally.

8.1.5   Triggers

In relation to triggers let me simply say that because of the presumption of both cause & effect, and the absence of spontaneity or randomness, any event based in Matter/Energy will require a trigger to generate it, however it is also necessary that **those triggers must pre-exist the event itself**.

This poses difficulties when we consider the issue of prime/ultimate origin, because it would seem impossible for any trigger to exist before existence itself.

Even if there was more than one underlying type of stuff in existence today, nothing could exist before existence.

So in the context above, the only way for the Big Bang to be explained entirely by physical means within causality, is to say that some form of existence has been eternal. Was that Matter/Energy, or the stuff of Thought, both of them, or even other factors?

The only alternative is to suppose that existence was a spontaneous event – without cause – and that is something which has traditionally been resisted by many scientists.

Matter/Energy is not believed to be capable of spontaneous activity.

8.1.6 Summary of Strategic Options

If there was truly nothing before the Big Bang then literally everything was created at that point, in the purest sense. On the other hand, if all matter/energy was already in existence and was simply re-shaped, then very little would need to be generated by pure creation – perhaps nothing at all.

Between the two extremes of 'all' or 'nothing', the core *types* of physical thing that might need to be <u>created</u> remain a refreshingly small list:

- a) the underlying essence of physical Matter/Energy itself;
- b) Forces including *Fields of Influence*;
- c) Processes and repeatable sequences;
- d) One or more Triggers;
- e) Configuration settings
- f) a Framework for Existence
- g) Time, and
- h) A source of spontaneity & randomness such as the underlying stuff of Thought

If these 8 factors represent the possible 'essence of existence' then those which did not emerge from pure creation would either have to evolve from the ones that were created, or they would represent the things that had no beginning.

They also represent the things which would occupy space, however empty space may extend in many directions. There is some debate about whether the number of dimensions is a base fact of existence, or whether new dimensions could be created. I have assumed that they are a base fact, however the number of dimensions which may exist is a moot point, as we can see in other chapters.

9   Scientific Fringe Areas

Having set-out our basic stall of scientific knowledge concerning physical reality and force, it's now time to extend the debate into fringe areas that have a surprising influence on our thinking about origin.

9.1.1   Nothingness, Antimatter and the Underlying Essence of Things

You would think that the concept of 'nothingness' is straightforward. It represents the absence of everything. However there are many debates which seek to alter this basic concept so that nothingness becomes something.

The reason for this is because it's another way to argue that pure creation is not necessary – as pure creation contravenes a fundamental and long-standing scientific proposition - that 'the equation always has to be balanced'.

Put more simply, mainstream scientific thinking doesn't do pure creation, but it is challenged by its own conclusions over the Big Bang if there had been no prior existence to provide a trigger.

There are 3 basic arguments to re-position nothingness:

1)   Nothingness is actually something – the 'zero form' of matter and antimatter.

2)   Force/Energy can exist in the emptiness of space without being instructed to take shape, (as physical matter), and in that formless state it can be argued that nothing physical exists but there might still be something present.

3)   If everything is just an illusion, without substance, then it is all in the mind anyway and nothing has to be physically created.

The original concept of Antimatter was that with enough energy, nothingness could be split into two parts of equal mass; one being the perfect 'anti-particle' to its twin particle of 'positive' matter. If the two came together again they would annihilate each other and return to the 'zero state'. People say this explains why some physicists won't eat pasta and antipasti in the same meal... but I'm not so sure.

The Antimatter theory emerged when equipment was less sensitive than today and 'Antimatter' was seen to appear as if from nowhere in very high energy situations, (such as cosmic rays hitting our atmosphere). After a fraction of a second particles of Matter and Antimatter would come together and simply disappear, so people began to think that the observed state of 'nothingness' might represent the underlying 'stuff' of reality. (This is reminiscent of the Tao in Chinese philosophy). In other words, empty space may in fact be 'something': a 'zero' state that couldn't be detected.

> Later, it was discovered that in very high energy situations such as particle colliders, two sub-atomic particles which had no mass, (eg. photons), could be used to generate twin particles which *did* have mass (such as an electron and its antimatter equivalent, a positron). Antimatter was no longer perceived to emerge out of nothingness: it simply combined existing particles and energy in a new form, (ie. empty shells of shaped forces being able to capture energy in order to gain mass).

The particles which we <u>now</u> label as Antimatter have been generated, captured and studied by science, showing that for every particle of matter there seems to be an equivalent antimatter particle of the same mass *but with the opposite charge and generally the opposite spin.* **There is no negative mass.** (Non-scientists needn't worry about the implications of this other than to say that opposite charge and spin do not represent fundamentally new elements of reality).

The other thing that distinguishes the new concept of matter and antimatter from the old, is that when opposite particles collide they ***do not simply disappear*** and return to a zero state as was first thought[35]; they destroy each other in a blaze of energy, light, and other smaller particles.

It has yet to be seen whether the original concept of antimatter can be resurrected in science through new discoveries, however the old 'zero state' concept is no longer pursued by science.

Yet the old principle is still used in debates about origin to illustrate a notion – that nothingness could be something. This has two purposes : it argues that

- an act of creation, (something out of nothing), would actually be a transformation event, allowing 'the equation' to balance.
- only in the 'magic' scenario of spontaneous 'pure creation', (ie. out of *absolutely* nothing), would a spontaneous act be required.

Whichever concept of antimatter we wish to pursue, there's another fundamental question: why, in normal life, do we only see matter and not antimatter if both are viable and seem to be produced in equal proportions at the same moment?

There are suggestions that antimatter is less stable and can break down into pure energy more easily than positive matter, while others suggest that antimatter has simply migrated to other parts of the universe, (such as the suggestion that vast clouds of antimatter may have accumulated around the black holes[36] at the centre of each galaxy). Science has yet to get firm answers on this.

9.1.2   The Mechanics of Time

As we have seen, Time has a pivotal place in theories about creation & God as it represents one of the possible prime factors in the origin of existence. In this Third stage of our consideration of Time I intend to look at its capabilities and some of the surprises they contain.

One of the key factors about Time within scientific thinking is that we have no absolute reference point against which we can measure it. We can only get a general idea about the progress of Time by comparing our circumstances to what is happening elsewhere. For instance we measure time in relation to the Earth turning once on its axis at its current speed - ie. a day.

Yet because all of our yardsticks are 'moving feasts' we cannot look at anything in isolation: we have to interpret what Time is doing to the things we compare against as much as understanding what is happening to us. This is the essence of 'relativity'.

> For instance: Time doesn't stop if we sit still. We might sit motionless for an hour and yet be aware that an hour has passed, because we have not in fact been still. Every atom in our bodies has been moving, and chemically reacting. We have also been ageing and deteriorating, a factor that would presumably only stop if every atom in our body stopped moving completely.

However science goes further, to suggest that the effects of Time can be different on objects which are travelling at different speeds. This comes from the equations which

Einstein wrote to model Energy use, which show that Time may slow down for objects which are moving faster.

This may seem fanciful, but as we'll see later, experiments have shown that this may be true in reality as well.

If correct, people and things would literally age more slowly if they were travelling fast, although it should be said that in order to produce a significant effect on Time an object would need to travel close to the speed of light. Not an easy task.

> To give an example, if a spaceship travelled at extremely fast speed away from Earth and then returned to Earth as fast as it could, when it arrived back, it's believed that people who had remained on Earth would have aged a lot more than the astronauts. The difference would become more significant the faster that the astronauts were travelling and the longer the time which the speed was applied (ie. distance travelled).

> So, plucking figures out of the air, the astronauts may have aged ten years while their compatriots left on Earth would have died of old age. The astronauts themselves would not perceive Time to have been any different on their spaceship even if Time was progressing more slowly for them compared to elsewhere.

> This isn't really time travel as popularised in novels and movies. If anything, it would represent time shifting – putting people into a later set of circumstances.

> In physical terms this would presumably mean that all of the atoms on the spaceship, (including those making up the astronauts) would chemically react with each other more slowly, so things would age or chemically evolve more slowly. However the experience for those on board would seem normal.

Although there are very significant limits on our technology, (ie. not being able to generate such high speeds, or anything close to it), there **has** been a test to see if Time does vary in the manner described.

> An experiment was conducted to see if this was truly possible by using a matched pair of atomic clocks that were perfectly in sync with each other. One was sent around the world on a jet airliner, after which it's time was compared to the other clock which had remained at the airport. It was found that the travelling clock had slowed by a fraction of a second compared to the one which had remained stationary.

So there is some evidence to support the theory behind the equations, and Time may indeed apply itself in different ways in different circumstances, but we should remember that these tests were very basic, produced the slightest of results, and were only conducted once, (as far as I'm aware).

In broad terms we can rationalise the theory in the following way : we have to pump much more energy into an object to make it go faster, and as we approach the speed of light truly vast amounts of additional energy are required to make speeds increase. In this way, while atoms are propelled across space, the huge additional amounts of energy would also add mass to every part of every atom causing their internal operations to slow down, making chemicals and people react less quickly, and thereby slowing down the rate of change (Time), while appearing normal.

An alternative 'reverse perspective' might be that the normal way in which the underlying force of Time operates may actually be to slow things down :

> Going faster may allow an object to 'collect' more Time than normal, thereby slowing its rate of change by even more.

If either notion is correct, Time will be working at a different pace for different objects in the Universe travelling at different speeds - which would cast doubt on the idea of Absolute Time – ie. an underlying 'set pace for Time'.

I also need to elaborate on an earlier point. We have to distinguish

- the notion of *changing the speed* of Time, (which would make activities in the current moment happen faster or slower), **from**

- the idea of *Time travel,* (where we generally suppose that people are taken outside Time and then reinserted into a timeline at an earlier or later point).

The implications of this will have significance later when we consider possible sources of randomness and spontaneity, however we first need to understand some of the other basic concepts surrounding Time.

It's worth noting here that while there may be ways for people to live in a future that they would not normally expect to be part of, (per the spaceship example above), there is no apparent way in which somebody could go backwards in time. We may be able to slow Time down, but it would always move forwards, not backwards.

However it is possible to *see* the past to some degree. Despite the incredible speeds at which light travels, the vast distances in space mean that it can take thousands, millions, and even billions of years for light to travel from its point of origin to another galaxy... say to ourselves.

In terms of the most distant galaxies, science tells us that what we see happing to them today may have actually occurred as long as 13.2 billion years ago. It would have taken that long for their light to reach us.

In this case, ancient light gives us a window to view the past, but not the ability to change it.

Finally, we have to consider what Time might contribute to creation if it did pre-exist the Big Bang. Could it be that Time was the only thing to exist beforehand, and if so, could Time itself spark creation? That depends on your concept of Time and the environment. If Time is a force in its own right then one might speculate that this force could

- be shaped into a physical structure (as discussed earlier), or
- tear nothingness apart to generate Matter and Antimatter.

However these are not popular perceptions of Time. People do not normally consider that Time itself is able to create anything. Indeed, it's generally believed that Time only enables things to change themselves, so the idea that Time is a force which can tear things apart is less persuasive as a result.

9.1.3   Dark Energy, Dark Matter, and Multiverse Theory

In this section we return to the question of the size of the Universe, looking for clues to explain why its expansion seems to be accelerating.

On the basis that the gravity of every object affects all of its neighbours and vice versa, one of the things that's puzzled science for some time is the behaviour of some planets, stars, and even galaxies which **can't** be explained by the visible material within and around them, (as observed with various types of telescope). Their orbits/movements suggest that there are other gravitational effects at play from objects/material that we can't see, and this led to the notion of 'Dark Matter'.

> An alternative may be to suppose that these stars/galaxies are evidence of a different type of matter, (configured differently to our own physical environment and having different properties)... which, it seems, people are **not** generally prepared to do. So we have to consider 'Dark Matter'.

Conceptually Dark Matter represents physical material which we cannot see because it is basically regarded as transparent. It doesn't emit/reflect/absorb any light and we haven't observed this dark stuff moving into our line of sight to 'block the view'. While the term Dark Matter manages to avoid unwanted connotations from phrases such as 'Invisible Matter' or 'Transparent Matter' we have to ask if this stuff is real?

The thing that's given us greater confidence in the concept of Dark Matter is a technique which can reveal distortions in the light that reaches us. If you imagine looking at an object at the bottom of a pool, the image will shimmer, and you may even see the light forming lines or rings where ripples run across the surface of the water. The light has been bent by the water, and in a similar manner it's believed that Dark Matter can bend light passing through it.

Observations on tiny patches of sky over several nights have apparently revealed such 'halo' effects when the different images are overlaid, and using this technique, maps of the night sky have been drawn up to broadly reveal where patches of this dark matter seem to be. Scientists have found that there can be many 'layers' of this stuff across the vast distances of the universe.

Unfortunately, matter that gets in the way of light and bends or distorts its path is not only likely to change the angles by an unspecified degree, but might also conceivably reduce the brightness of the light reaching us even fractionally, so how could we measure its effects on luminosity? In truth, we have no basis on which to determine the degree to which this is happening because we have no idea how transparent this Dark Matter is... or indeed whether it has any effect on luminosity at all.

If you remember the prime techniques being used in astronomy of 'Parallax' and 'Luminosity' (Chapter 4.1.2) such distortions could radically alter our view of the size of the Universe. Conceivably it might even bring our estimates of the size of the Universe within 27.4 billion light years - avoiding any need for the theory of Inflation.

However that is pure speculation so if we stay on topic, the theory of Dark Matter has implications for the expansion of the Universe because it should increase the effects of gravity within a region of space, (a force of attraction), which should slow the expansion down even more... yet the reverse is happening: the rate of expansion is accelerating. Why?

In short, nobody knows, so science has invented another concept: that of 'Dark Energy'. This is also invisible to us, but it's intended to work in the opposite way to

Dark Matter by pushing the Universe outwards. To 'balance equations with our observations', it was originally calculated that Dark Energy must amount to 73% of all matter/energy in the Universe - none of which we can see or detect in any way.

As you will see opposite, thanks again to the WMAP survey, NASA has refined its estimates down to 71.4% but this still means that 'Dark Energy' must be virtually everywhere, including possibly here on Earth.

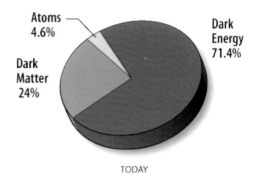

(Credit: NASA/WMAP Science Team)

Physical Matter and visible material (atoms) represent 4.6% - quite a small proportion of the total, yet in many respects the Dark Energy figure seems to be more of a 'balancing factor'.

If we remember the results of the Dual Slit experiments and the effects which are reminiscent of waves being generated in a pool of hidden stuff, then Dark Energy could easily be that other stuff... if it really exists. All of this becomes significant in 9.1.4 below.

Unlike Dark Matter, the theory of Dark Energy is mainly speculation/hypothesis, although there's a growing belief that the 'Late Time Sachs–Wolfe Effect' may be a pointer to its presence. In broad terms this argues that some regions of space have concentrations or 'wells' of Dark Energy, and there are other places where Dark Energy is absent. The wells would add energy to any light passing through it, while voids would reduce the energy in light. That at least is the speculation.

Yet the accelerating expansion of the Universe has deeper implications because it is believed by scientists to have started 8.77 billion years after the Big Bang[41], (just before the emergence of our solar system). This is very significant to our debate, because it means that the force which has been driving the acceleration for 5 billion years was probably a *new* force, with a new source **that must continue to generate energy for as long as the acceleration lasts.** Where did all of this Dark Energy come from? It must represent a significant and prolonged event to generate that much stuff.

> We should remember that when we accelerate a car we burn more fuel.
> For as long as we push the pedal down and wish to continue accelerating,
> our petrol/gas will be used up to provide the energy to accelerate, and
> when we stop accelerating the car will continue to move at a steady but
> higher speed until something else slows it down.
>
> In broad terms the same should be true of the Universe.

In short, all current explanations are on the basis of speculation, however it is important to consider the broad options when the implications are potentially so profound. Was some energy immediately 'archived' after the Big Bang and it is this which is now being released? Is there new source of energy beyond our 3D universe? Is this the result of pure creation? Put another way, we are again forced to consider whether a spontaneous or **ex**ternal influence may be responsible.

Apart from sources of energy we have to ask what can actually do the pushing?

If we imagine the Universe at the age of 8.77 billion years being a spread out set of galaxies with very little in between, then there would seem to be insufficient material to do the pushing. The stars within a galaxy are not rigidly tied together in the same way as the components of a car. They are floating in space and even if we say that one of these could explode to create a pressure wave as its material spread out, the largest explosions we ever see are supernovae and these are not powerful enough to cause an entire galaxy to accelerate outwards or even to make it swell a bit.

In other words, Galaxies are not like a car or plane which you can push with a supernova engine, even if you directed the blast of a supernova in one direction, (like a jet). Something much bigger than the galaxy would have to push every particle within it at the same time with the same force if the galaxy wasn't to become squashed... (and of course we don't see such distortions). Neither have we detected any pressure wave that's spreading outwards causing galaxies to bunch up as if they were affected by a sudden huge blast.

In short, there doesn't seem to be a normal chemical/nuclear process to explain this expansion, and even less to say why it was suddenly triggered 5 billion years ago as a sustained release of energy. So how might we explain it?

Ideas about Dark Energy from Within the Physical Universe

If Dark Energy was part of the original material of the Big Bang, (part of the original singularity before it exploded), it's hard to see that it could represent something new or *additional*. We would expect this energy to form part of the power of the initial blast, which drove the material outwards at maximum speed. We wouldn't expect this stuff to *remain as energy* but not used, (ie. preventing its effects from being felt). Archiving energy would seem to require a change of form.

As most scientists believe the acceleration effect is real and that it requires energy to power it, there only seem to be 3 possible source, (which I briefly mentioned earlier):

- energy was converted into a form of physical matter as a means of holding it in reserve, and that material , is being returned to an *active* form of energy, **or**
- there is an external source of new energy,   **or**
- new energy is being created from nowhere – pure creation.

As we can't measure Dark Energy, (or even detect it), there is no way of knowing which of these possibilities might be correct – if it exists at all. What we can say is that there is no evidence for huge amounts of normal physical material being converted back into energy. The odd supernova or the energy released from black holes doesn't seem to be anything like enough to produce such a universal effect.

We may therefore speculate that an invisible process may be converting Dark Matter into Dark Energy as a prolonged process, (on the basis that both of these things are also invisible). This thinking is the prime way in which to maintain the doctrine that our Universe is the totality of everything.

It could be argued, for instance, that if Dark Matter exists everywhere, the energy released could be applied everywhere at the same time. The counter argument is that Dark Matter seems to exist in patches across the Universe, and there is no evidence of this stuff converting into energy as a sustained process in the amounts we might imagine, or with the high level of co-ordination that would seem necessary.

Without an inevitable process that would cause this timely effect we come back to basics to say that a spontaneous or **ex**ternal influence could well be required.

As we saw in the last chapter a very different alternative is to suggest that the simple presence of Matter/Energy may distort the 'fabric of space', in a way that would push material outwards towards a void. This uses the same notion that was used to explain gravity, but producing the opposite effect – a force of repulsion not attraction.

> In visual terms if we remember the 'sheet of elastic' explanation of gravity, this fabric of space would cause a hill instead of a dip when it encountered a physical object, as an analogy of why other objects on the slope would move away. This could explain why 'nature abhors a vacuum' and why matter always seeks to fill it.
>
> This would represent a 'structural' explanation rather than one based on chemical/nuclear reactions, but it would also mean that Gravity would have to be explained by other means, (eg. Stephen Hawking's suggestion that it represents 'negative energy' – which we will see later).

Clearly this relies on there being a 'fabric of space' or an alternate influence similar to gravity which we haven't yet detected or identified.

However the really challenging fact for this theory is that such material would always be present after the Big Bang – its effects wouldn't kick-in after 8.77 billion years. The way that has been suggested to overcome this objection is that the effect was always present but gaining in strength and it was only after 8.77 billion years that it became strong enough to overwhelm the slowing effects of gravity etc.

While there are equations to demonstrate how this might work in practise, the truth is that these are encoded speculation, and we each have to consider whether one idea is more likely than the alternatives.

Peoples' attention therefore turns to **ex**ternal sources – which could take many forms.

Ideas about Dark Energy from Outside the Physical Universe

If we imagine that our physical universe is expanding into an empty space, (which was much larger that the Big Bang and potentially infinite), it is likely that the 'void beyond' will have always been there and one way to imagine that it has been exerting an influence is by generating a **new** type of 'suction force'. That force **would** represent something additional to Big Bang explosion but it is also likely to have been present from the first moments after the event, so this again struggles to explain the sudden acceleration that emerged 5 billion years ago.

We need to consider other explanations.

If you think that there are other dimensions of existence beyond the 3 that we occupy, (as proposed by M-Theory), then these may hold energy and material in the same 3D locality as our Universe but within those other dimensions and therefore beyond our ability to detect it – an **ex**ternal influence. However to sustain an acceleration of the Universe like this, energy must be continuously flowing into our dimensions across the full region of our Universe – helping to push everything outwards in some way.

The starting of this flow of this energy would suggest that either the energy was suddenly made available from the other dimensions, (a hole emerged), or that the 'barrier' which normally keeps the dimensions apart suddenly became porous or more porous. The advantage of additional dimensions is that they would not be truly **ex**ternal to the Universe, only **ex**ternal to our 3D space.

Either way, it is possible that an inevitable process might bring this flow about instead of a spontaneous influence, and this is significant when we think that explanations within our 3 dimensions have to comply with the Laws of Physics as we know them. In other words, by having a very broad effect across the whole Universe explanations are not challenged by the limits of the speed of light.

A final set of options to explain the accelerating outward movement of galaxies says that our Universe is **not** the totality of everything, by arguing that other things lie beyond it. Within our 3 dimensions this additional material can be thought of as:-

- multiple universes effectively sitting side by side within our 3 dimensions, and/or

- more chaotic or unstructured material that is scattered everywhere across an infinite existence.

As we first considered in Chapter 4, the suggestion is that these other types of material would be exerting a ***gravitational*** influence on us – pulling us outwards in all directions & causing the acceleration.

Yet this notion also has its problems, because it would mean that this **ex**ternal gravity was being applied perfectly evenly all around our Universe - which could be quite hard to achieve – especially if there were other large blobs of existence (universes). On the positive side, it would suggest that the galaxies closest to that external material would be most affected by its gravity, (causing them to move outwards faster), which does fit the pattern of movement that we see. The challenge is in explaining how the acceleration began 5 billion years ago.

While it could be argued that this external material is approaching us, and therefore its influence only became noticeable 5 billion years ago, the idea that these objects are approaching us from all directions and in a perfectly even spread seems highly unlikely, (both in terms of distances and density). You can hopefully see why multiverse theory is failing to persuade many people that it can explain the accelerating expansion of our Universe.

We will see later why the possibility of Parallel Universes, (in other dimensions as touched-on above), might also represent a good explanation of the Big Bang and therefore a single solution to both criteria.

9.1.4 Multiverse Theory

Although they struggle to explain Dark Energy, other Universes remain a serious consideration even if there's no direct evidence for them, (ie. no overlapping circles of expanding material, or dents/bulges in the edge of our Universe to indicate a collision/interaction). There are many 'Multiverse' theories, and the term was first coined by American philosopher William James in 1895.

However the cosmologist Max Tegmark did present a taxonomy/classification of universes, and beyond this Hugh Everett is renowned for his 'Many Worlds' interpretation of quantum mechanics. In short the concept is taken seriously.

One of the ways in which it can be argued that there <u>have</u> to be many Universes is the notion that if inflation occurred there may be nothing to stop it, so even if the material we can see is travelling below the speed of light, the outer material of the Big Bang explosion might still be expanding the overall size of the universe to this day, at the speed of cosmic inflation and to near infinite levels.

Alternatively it can be suggested that if existence is eternal and infinite then if a Big Bang event can occur here it may also occur in a myriad of other places too.

On one or other basis, it can be argued that other groupings of matter may exist similar in size to our visible universe but beyond our view. However playing with infinity like this involves a lot of speculation.

Despite the lack of evidence, (other than mathematical probabilities), the reason why Multiverse concepts are seen as desirable is that they "allow nature to experiment" with every permutation/configuration of existence more rapidly. This is a prime way in which to avoid luck in generating our configuration of Universe - the one where nature got the circumstances right. Yet I say again that there's currently no physical evidence for any other Universe than our own. It's all theory/speculation by scientists.

9.1.5   M-Theory & Implications from the 'Faster Than Light' findings

It has been a strong and long-standing principle within science that nothing should be able to exceed the speed of light, and yet we have seen various measurements and experimental findings which now show this to be a possibility – and not just by a marginal effect, but by a multiple of many thousands.

This is true in terms of both physical movement, (because of the supposed effects of Cosmic Inflation), and instant communications/responsiveness, (due to the influence that has been demonstrated between paired sub-atomic particles such as photons and electrons travelling apart at twice the speed of light [19, 32]).

Put another way, scientists can't dismiss these factors as being non-physical and possibly **just** related to data. We have already considered explanations based on 'other dimensions' which have their strengths and weaknesses, but I can think of other ideas that might help to rationalise these findings within the Laws of Physics.

The theory of Inflation suggests that in the earliest moments of the Universe after the Big Bang, the underlying stuff of Matter/Energy expanded far faster than light before slowing down again to speeds lower than the speed of light.

One concept for how this might occur uses the mechanisms which might shape the basic particles of existence to create more occupiable space, as follows:-

String Theory[21] and its enhanced version 'M' Theory[22] have suggested that microscopic forces can be shaped to provide the basic structures/shells of sub-atomic particles. Spheres and cylinders have been suggested as possibilities.

Conceptually, before they were shaped, 'sheets of force' would have lain flat, squashed against each other in the compressed state of a singularity, and yet occupying little or no depth as they would have no real thickness. We can liken the situation to a ream of 500 sheets of paper, (although this actually has a depth of 2.25 inches, 5.5 cm). So let's run with this scenario.

Let's say that each of these sheets is A4 size and is instructed to take shape, causing its short side to bend round to say create a tube/cylinder measuring 10cm or 4inches across the width/circular end. They would push all other sheets that were next to them out of the way in order to make space for themselves. So if all 500 sheets were suddenly instructed to take shape, we might find that the depth of that ream would now become 50 metres / 165 feet wide – close to a thousand times wider than before.

If the 'instruction to take shape' was transmitted through the sheets at the speed of light, and each sheet managed to adopt its new shape virtually instantly with no distortion or cushioning effect, then the speed of expansion for the whole 500 sheets would occur at a thousand times the speed of light, in terms of distance. Instant communications would raise this even higher.

If such thinking is correct, it would be the instant shaping of all sheets that would explain the incredible rate of expansion perceived by the theory of Cosmic Inflation, and also explain why it suddenly stopped, (all active elements within the singularity having taken shape), which is equally important to the overall concept.

For those interested in the minutiae of this logic, the outcome can be radically altered depending on:

i) whether the instruction to take shape was communicated instantly across all sheets; or was transmitted by an unknown vehicle (say a particle of light cutting through the sheets); or was passed by contact from one sheet to another at a certain speed.

ii) whether there was a delay between an instruction being received and the force taking its final shape, (which might allow the instruction to move ahead of the expansion and not be overtaken by the expansion), or whether the lag was of such a period that it slowed-down the overall expansion of the Universe.

This is one way to envisage how Cosmic Inflation might occur within our 3D space and within the known Laws of Physics, (given that faster-than-light communications appear to be a reality). Yet, we don't have anything to prove that M-Theory is correct.

Equally, we don't know what allows paired particles to communicate so quickly **today** rather than in the first moments of the Universe - thereby breaking the Laws of Physics, (although people inevitably speculate that it is something related to additional dimensions of existence). This is where ideas about additional dimensions of existence could help to bridge the gap... although in true this is just a vague idea with no specific explanation.

We can tie the two ideas together by suggesting that (say) a 2D existence was suddenly increased by a 3rd dimension which required existing material to expand into its new space and take shape. However this skates over one awkward point – that we have no concept of how an addition dimension could suddenly be added to existence.

The only other concepts that I am aware of which might rationalise Cosmic Inflation suggest that if material crosses from a hidden set of dimensions into our 3D space

- at a particular location, then it would almost certainly force that material to be reconfigured into our format. As part of this it would have to shed its old configuration, (laws of existence), and in that moment of transition it might be free of all Laws of Physics – enabling it to travel faster than the speed of light until the new laws of physics were applied.

- across a broad area of space (say 70+billion light years), then there would be no need for cosmic inflation – everything could travel at below-light speeds.

10 Scenarios for Origin (excl. Thought/Spontaneity)

If you talk to people about a huge and dramatic explosion which could explain the origin of their own existence you're likely to find a ready audience because it will appeal to the little boy in most men, (including me), and probably to a large number of women too. It's exciting... but it's not the only explanation.

The reasons why we keep looking for explanations are threefold. Firstly, all current explanations have their flaws, and those flaws need to be eliminated. Secondly, facts and their contexts can be interpreted in different ways. Finally, new findings can have a dramatic impact on the validity of earlier theories. The imagination displayed in trying to resolve these points can also be great fun!

This chapter will cover the range of current thinking about how our universe may have been created if it **wasn't** the result of an exploding singularity, but still only using the perceived capabilities of Matter/Energy. The next chapter will build on these ideas to see if any broader possibilities arise by including 'other types of stuff'.

Whether you judge the ideas to be viable or not, you should find them interesting and entertaining, while also pointing to clever notions that we may find useful in future. I say again that **no** current explanation is without significant problems, so even if the concepts sound strange at first I would ask you to give them all a fair hearing.

As you can imagine, when creating something as complex as the universe there are a few angles to cover, so we'll have to work at it a little bit, but in the end we'll identify the more realistic choices that are available, based on current knowledge and concepts.

Explanations of origin need a structure and, as with any process, there are 3 basic components to such theories:- a **starting position**; a **mechanism**; and a **set of outputs**.

In terms of starting point, the philosophical logic of prime origin is surprisingly simple in only offering 3 basic options:

a) If you start off with a 'blank sheet' (ie. a total absence of anything), then you need something to be created out of nothing, (*spontaneous pure creation*)

b) Existence may have no beginning, (ie. it is eternal), and therefore any change that we see is a *transformation* of what was there before, but nothing is added to, or taken away from the overall pot of existence.

c) *A combination of the two* – where pure creation adds to any existence which had no beginning.

Surprisingly no theory that I have ever seen suggests that something could be *removed* from the 'pot of existence' although that remains a logical possibility. The assumption seems to be that if some 'theoretical stuff' doesn't exist now in our stable environment, then it wasn't necessary. Yet even if something was needed to either stabilise nature, or facilitate its transformation into this form, we have no clue what it might have been.

But I digress... as part of the overall concept, we also have a lot of clarity about the *outputs* which must be achieved by any theory of origin. This reflects the items discussed in Ch 8 however we can also say that the outcome must lead us to an **exact** match with the cosmos as observed today. In this respect 4 key findings are pivotal:

- **The Universe is huge**. Even the bits that we **can** see extend far beyond any explosion that was travelling at the speed of light, (the maximum possible speed as determined by science).

- **'Stuff' is very evenly spread out**. While the visible material inside the Universe is clumped into galaxies, there are vast numbers of them which are spread very evenly across space - which is difficult to achieve. It also implies that there were no major overlapping creation events, or collisions with other universes.

- **Pattern of movement**: with one exception all galaxies seem to be moving away from us, and the more distant they are, the faster they seem to be travelling. In addition, the *increasing* redshift implies that their speed is also increasing, not slowing down, (ie. the expansion of the Universe is accelerating).

- **One version of physical reality**. The Universe seems to apply the same rules of existence everywhere.

At first glance you might think that these would be fairly simple criteria to meet, and yet so many ideas struggle to match them, even though peoples' imaginations are allowed to run free. We therefore find ourselves hanging onto the few ideas that get us close to our goal, even if most of them sound a bit far-fetched at first hearing.

Strategically, new theories need to flex the process of origin to present us with a different approach, and if both the starting position and the outcome offer little room for manoeuvre, it becomes easy to see why the greatest scope for new ideas lies with the ***mechanisms*** that may have been used.

Across a number of theories various mechanisms have been suggested. Some of these are based on the *known* characteristics of matter/energy, while others are more speculative. We each have to judge the merits of any suggestion on the likelihood that its mechanism will exist and will also succeed, (including whether it relies on luck to produce the correct result). In addition, theories must also match all of the facts which we have accumulated through science, etc. if they are to be seen as viable.

As you will realize by now, no theory can currently do this but some can come a lot closer than others. Equally, some mechanisms can work with some starting conditions but not others. So a part of any person's assessment must be how much faith is required to fill the gaps – whether that is from a secular, atheistic, or religious perspective.

The theory of an exploding singularity was originally accepted because its suggestion of an explosion did seem to match the pattern of movement in the cosmos, while also using a known feature of matter/energy - explosions. It also worked with **all** of the philosophical options relating to starting conditions 'a-c', (because it could be applied to either a moment of pure creation, or a transformation event).

Unfortunately, it was later considered potentially flawed for 2 reasons. Firstly, it could only work by breaking the known Laws of Physics (re: the size of the Universe), and secondly because the explosion is not slowing down but apparently getting stronger. Sadly, the main way in which these challenges have been 'explained' is to suggest that

- 'Big Bang' events are deemed to be the only things that are able to break the Laws of Physics in order to go faster than the speed of light,   plus...

- there is an invisible component in the Universe, (Dark Energy), accounting for the accelerating expansion. This new stuff would have to represent over 15 times the amount physical material we can see in the Universe (galaxies), if it was to balance the relevant equations, and yet we cannot detect it.

Unfortunately there are at least two other conceptual problems with this theory. Firstly, there are severe doubts that a singularity would be guaranteed to explode in the ways expected of it, on a regular basis for all eternity, (as we will see later).

Secondly, as a one-off event an exploding singularity couldn't explain how our particular configuration was achieved without luck. This is where the original idea was modified to reflect a series of Big Bang events, (Bang-Crunch), but this has also been undermined by the accelerating expansion. The cycle must have ended, and a finite process can't guarantee a particular outcome.

You can hopefully appreciate why the bulleted responses above are failing to persuade everyone, and why the appeal of some new theories is growing. Equally, as Materialist concepts no longer have the evidence to prove their case, they can no longer use evidence to claim superiority over other theories. We will find much greater significance in the *number of problems* which each theory contains.

We will run through the key scenarios in a moment, and aside from finding them interesting I hope you will see the potential within them, even if they are far from a complete explanation. Those with the least number of problems may be our best pointers to the truth of what happened. For the sake of completeness, I have filled some of the logical gaps in the range of possibilities with some suggestions.

While this may sound a little abstract and dry, you haven't heard the theories yet, and they don't have to be boring. So to ease us into the analysis, perhaps a little milk and sugar might help.

I say that because you can draw remarkable parallels between the Big Bang and making porridge. Both seem to suggest that out of a hot and fluid mixture, solid matter can materialise.

The basic analogy requires: a pan, (a framework for existence); plus the water and oats (equivalent to the underlying stuff of matter/energy); heat (the power to transform); something to stir the mix; and possibly something to flavour the end result - adding the milk & sugar... or salt !

It's a way to remind ourselves how significant some of these abstract ideas are. So...

How do you make a pan without metal or any other material? How can you produce oats if there's no soil or seed? In these examples the principles and significance of pure creation are clear. Pans and soil don't normally materialise out of nothing – so how can we ever expect this to occur? The potential of spontaneous pure creation is covered in the next chapter because such an event would require other capabilities beyond those we observe in Matter/Energy.

By implication therefore, **this chapter must assume that everything has evolved from things which have existed forever**.

We will return to the porridge scenario in a little while, because I first want to equip you with ways to assess the viability of ideas.

10.1.1   A Toolset for the Assessment of Ideas

Let's imagine that you're going to write a recipe book on how to make various types of universe. Where would you start? Well I'd begin with a list of available ingredients that might assist us... but as we can't see what was there 13.77 billion years ago, that becomes a matter of speculation.

Worse still, we should remember that we're talking about things which would need to exist <u>before</u> the Big Bang, and if literally everything was fundamentally changed by this event, nothing that exists today can be expected to resemble any prior existence. So our ingredients can only be described using generic factors.

In short, while Chapters 4-9 presented a lot of evidence to describe the nature of existence today, the topic of origin has virtually no evidence - hence the separation.

For this reason we really do have to speculate on what ingredients may have been available for our recipe, and **the good news is that *you* have the power and ability to dream up whatever you want**... so long as it's credible and guaranteed to produce the end result – our universe.

So in this debate you not only get to 'bake the cake' but also get to judge the results of the competition. Well, you can't say that I'm not generous!

Your shopping list will no doubt be based on the concepts about origin which you believe to be credible, and because this section excludes the possibility of God in order to see what would have to be achieved 'the hard way', this will inevitably reflect scientific findings.

In addition, a number of criteria will determine the ***quality & viability*** of any proposed explanation for our origin. For instance we need to *avoid luck* in our explanation; must achieve *standardisation* across the Universe; explain any measure of *control* that seems to enforce the Laws of Physics; and explain how things can change.

We can therefore assemble a checklist of criteria to validate our list of the ingredients and mechanisms involved – a toolset with which to judge ideas about origin.

This will take the form of a number of principles, or logical requirements, that need to be observed if any explanation of origin is to retain its credibility. You should find the list to be quite reasonable if a little dry in its reading, but I will bring these criteria to life with an example at the end - including a conclusion that may surprise you.

As it <u>is</u> important to understand this logic so that I can short-cut some of the analysis later, can I ask you to please stay with it.... there are only 8 items in the list!

In relation to overall conditions:-

a) **'Pure Creation' must emerge from <u>absolutely nothing</u>** even if the outputs are instantly mixed with something that already existed.

b) The principle of causality means that **Matter/Energy is never thought to be able to change the rules**, only to live within them, yet there may be layers of reality in which different rules apply.

c) **To have an eternal cycle the same rules (Laws of Physics) must be applied each time around,** so some level of physical existence must always be preserved though a Big Bang, and at that base level matter/energy will rigidly conform to the same rules whether that is **before, during,** or **after** an event.

If things don't appear to conform then we have either misunderstood the rules or **something else** has changed the rules.

It has generally been assumed that within a singularity physical matter/energy will be reduced to its most basic form - the raw stuff of existence, and this will operate a base set of rules that might be applied in different ways to produce different configurations of existence at our level of reality.

Yet there is nothing to say why those different configurations might arise. Each time a singularity emerged, it would always be the exactly same material governed by the same set of basic rules that applied to the raw stuff of existence.

In relation to the management of luck :-

d) **Infinity should guarantee that eternal processes will achieve every possible outcome at some point**, avoiding the need for luck/chance when seeking a particular result, although the mechanism itself may limit the range of outcomes

However this principle is a double-edged sword:-

e) **All end points** mean that the activity in question is **finite, not infinite**, and therefore it cannot be assumed to explore every possibility. Most start points will also point to a finite process.

f) If a cyclical process cannot explore every outcome because it is time-limited, **(finite),** it will either require **luck/chance** to achieve a particular end result, or will require an **element of design**.

Finally, in relation to the need for spontaneity:-

g) **Any change to an eternal condition, eternal activity, or eternal process must involve an external or spontaneous cause** and probably luck as well. A closed system can only be changed through spontaneity/randomness.

h) To our knowledge, **the only factor in existence which might be spontaneous** (a source of change), while also avoiding luck in its outcomes, **is Thought**.

So these principles form our basic toolset. There may well be other principles that could be added to the list however I have found these to be the most useful.

Inevitably a list such as this seems very abstract and unmemorable without putting them to use, so to help you get into the spirit of how they might operate let's apply them to an example while they're still fresh in our minds:-

> The discovery that the expansion of the Universe is **accelerating**[17] came as a real surprise to many scientists and meant that the universe will not collapse again through the force of gravity. Yet it's the logic of the toolset which now leads us to additional conclusions.
>
> If there had been an eternal 'Bang-Crunch cycle' it would have been ended by the last Big Bang, (because there will be no 'crunch' at the end of this Universe), so the cycle would either have to be broken via a *spontaneous* change, and/or by an **ex**ternal factor, because physical matter isn't able to be spontaneous and can't 'change the rules', it can only live within them.

This isn't just of passing interest, it's very significant. It either means that the Bang-Crunch cycle:

- is invalid, or
- was finite and therefore incapable of exploring every option – thereby requiring luck or a thoughtful design to achieve our universe, or
- was previously infinite, but then subject to a spontaneous change, so in order to remain valid it now requires something to be the source of that spontaneity.

In other words, if a reliance on luck is unacceptable then **the Bang-Crunch cycle now requires the underlying stuff of Thought to make it work,** (as the only known source of spontaneity), or something external to physical existence.. like God.

A number of scientists have responded to this by setting aside the bang-crunch model – arguing that science had misunderstood what was happening.

They now propose that some other type of eternal cycle may exist *which would be unaffected by the continuing expansion of our Universe*, but would still allow Big Bang events to be generated on a regular basis.

The 'Curtains of Energy' from Ch.4 is one example of such an alternate mechanism. We'll take a closer look at this new theory shortly, but at the time of writing it has no evidence to suggest that these structures/curtains may actually exist. **None**.

This is different to the Bang-Crunch cycle because when it first came out it **did fit** all of the available evidence, but new scientific findings have now undermined it – leaving **all** viable atheistic models *without* real world evidence, as mentioned before.

So all views of origin, (scientific and religious), are now effectively based on faith.

10.2   <u>Some Lateral Thinking</u>

In this section I want to explore the *principles* behind a small number of ideas which will be useful later. They will involve some mental gymnastics but they unlock some of the detailed and interesting suggestions we will encounter shortly.

The first three sets of concepts look at ways to explain the shape, size, and dynamics of our Universe **without breaking the Laws of Physics**. By implication they must avoid notions of a central start point, (exploding singularity), in order to avoid physical material having to travel more than 13.77 light years in the time available, (ie. to bring the speed of travel below the speed of light).

They therefore begin by suggesting that a different sort of Big Bang event generated an effect on **material that was already widely spread-out**. This spread might be because physical existence has always existed everywhere, or it may be the product of cyclical mechanism like the 'curtains of energy' which may touch across a very broad region and not just at a single point.

The next stage of this thinking is to determine what type of event the Big Bang might be. The first set of ideas re-define it as a different type of explosive event, creating pressure or momentum from *within* the Universe.

The second set of ideas look at influences that may originate *outside* our universe – exploiting the potential for voids and vacuums in the infinity beyond.

The third set of ideas considers how our configuration of existence might be achieved without resorting to luck, using self balancing mechanisms, or Frameworks.

The fourth and final set of ideas take the alternate view - that it is **necessary** to break the Laws of Physics, not only to achieve the size of our Universe through the 'mechanism' of universe inflation, but also to allow the Universe to explore every permutation of existence as part of an eternal cycle, to avoid the need for luck. This type of thinking allows us to preserve notions of a central start point, whether that is an exploding singularity or some of the other mechanisms we will encounter.

Here the emphasis is on explaining **how** the Laws of Physics might be broken if physical material normally has to stay within the rules and not change them.

Once you appreciate the basic thinking behind these 4 sets of ideas, it's becomes a lot easier to see how they might be applied in more specific scenarios.

1 - Explosive Pressure on **Spread-Out** Material – within the Laws of Physics

While a *single* explosion remains a strong theory to explain

- the even distribution of galaxies across our Universe and
- the pattern of their movement,

the sheer size of the *visible* universe, (let alone the regions that we can't see), means that no explosion acting within the Laws of Physics could produce the spread of material that we observe in the time available since the Big Bang.

The laws of physics restrict the spread of material to a maximum of 2x13.77 = 27.54 billion light years - not the current estimate of 93+ Billion light years

Although it is conceivable that new factors may radically change our estimates of the size age of the Universe, science believes that its current estimates are supported by a number of different facts and therefore we should trust these findings, (which have been carefully assembled and checked), until there are new reasons not to.

The only way for a single event to stay within the Laws of Physics is to assume that a different sort of Big Bang acted upon material that was already widely spread out.

If Matter/Energy had existed in a dynamic state for all eternity before the Big Bang, it might seem perfectly natural for it to have been scattered far and wide. Conceptually the only factor that could have drawn material back into a compressed state/singularity is gravity and yet we **don't** see gravity acting in this way on our Universe – its expansion is accelerating.

So let's assume that the raw material of the Universe was indeed widely scattered immediately prior to the Big Bang event – say 70billion light years across. This would mean that after the Big Bang it could just manage to spread the remaining distance if it travelled slightly slower than the speed of light.

In these circumstances, what could cause the ball of the universe to expand at the same rate in all directions, while preserving an even spread of material within it? (An explosion in just one part would cause a bulge - thinning-out the material in that region). Logic suggests that there would have to be a dominant effect which caused the **whole** of the Universe to move at the same time. In the first set of scenarios the effect would be driven by **multiple** explosions spread across the Universe.

If this set of explosive events could be triggered at the same moment, then the potential exists for the associated blast waves to press against each other, collectively raising the 'pressure' within the universe to an even level. The only way for that pressure to be relieved would then be for material to collectively move outwards but from widely spread-out starting positions.

(Remember there may not have been any galaxies before the Big Bang, just an even spread of gas-like material everywhere).

You may well find this to be a very contrived notion, but it is just about conceivable in the right circumstances, and it's the *principle* we should be interested in.

There are two main scenarios for the 'many explosions' idea. The first suggests that there may have been one for each galaxy we see today, (ie. 200 billion 'explosions' evenly spread across billions of light years of material existence). The other suggests that every particle in existence was instructed to explode at the same moment. You will see how these ideas can be brought about later.

Explaining the 'galaxy level' approach is probably sufficient to illustrate the thinking.

Large isolated explosions in the midst of a space being occupied by gas or fine material would initially create a series of ball-shaped influences, that were expanding in all directions.

Like a myriad of inflating balloons filling a chamber, each explosion would produce some form of pressure wave which would eventually come into contact with its neighbours. While the general outward movement would emerge from the equalisation of such pressures across the universe we might suggest that this pressure would be relieved more quickly on the outer edges of the Universe, causing those galaxies to move outwards quicker... which is indeed what we observe.

With even more speculation we can even extend this idea to suggest that apart from causing the entire Universe to expand, when each galaxy-sized ball of pressure pushed against its neighbours a shock wave rebounded, pushing material back towards the middle of each ball compressing that material and creating a black hole at the centre. In this way each galaxy might become discreet as well as moving both outwards and away from its neighbours.

As the material of the Universe would have originally been widely spread out and none of the galaxies we can see are separated from their neighbours by distances that come anywhere near 13.77 billion light years we can be sure that such a mechanism could operate well within the speed of light.

While it's easy to see the appeal of such notions, many people will feel that the basic idea is fanciful at first hearing because it relies on a number of unexplained factors :-

- a cause for this type of Big Bang event.
- a way to simultaneously trigger many discreet explosions across vast distances.
- a type of reaction that could produce explosions on a galactic scale.
- starting conditions which could enable pressure waves to push against each other.

All of these points do need to be answered if associated theories are to be considered viable, and we will see that they do this with various degrees of success. For instance, the simultaneous triggering of explosive events across a scattered Universe could be orchestrated through the mechanisms implied by the Gisin experiments[19] where paired particles *instantly* mirrored each other's effects, even when separated by many miles and travelling apart at twice the speed of light. There would still be a need for a trigger.

> To be clear, we would **not** be saying that a myriad of paired particles had to be scattered across the universe to allow this communication to happen.

> The suggestion would be that the paired particles in those experiments were tapping-into a *general capability within nature* that could be utilised by something else to co-ordinate actions across the universe.

We can leave the notion at this stage because my intention is simply to demonstrate various underlying ideas so that I don't have to repeat them in each scenario later.

2 – Generating Voids within the Spread-Out Material of a Universe

At first glance you may think it very bizarre that the creation of a void could be a mechanism for the Big Bang when it implies that the material of the Universe may already exist. Yet a void can be used in two ways. The first is as a potential source of spontaneous pure creation, (discussed in the next chapter), while the second is a way to generate powerful forces that could cause Matter/Energy to explode outwards.

We know that energy naturally flows from hot areas to cold areas and that this effect will stop once the heat equalises between the two zones. In similar manner high pressure air will flow to areas of low air pressure to (say) to fill a vacuum, or create weather effects such as a tornado.

In both cases the power which drives the effect comes from the place which has the greater concentration of material, and it will stop when the zones are equalised.

On this basis the principle would suggest that if the Universe has a higher concentration of energy or matter than the void/vacuum which surrounds it, then there will be a tendency for material to flow from the Universe into the void and, depending on the difference in 'pressures', this might result in some very powerful flows.

Although the volume of material being moved might be very large, (possibly even on the scale of a Universe), we **wouldn't** expect this effect, by itself, to generate speeds that came anywhere near the speed of light. However as a basic notion it may help to explain the generation of a 'first singularity'.

If we look at the unbelievably huge distances between galaxies, and see nothing, it's tempting to think that there's a lot of complete emptiness in space, and therefore there would be no pressure difference to cause such a vacuum effect along these lines. But that's not necessarily correct, not only because the galaxies could have been much together in earlier days, but also because a number of things occupy all parts of space

From the cosmic background radiation, to particles of light, dark energy, and dark matter, empty space is far from empty. So it may actually be more realistic to say that a sizeable void in space could be very rare. There is also speculation that it may be a much more powerful thing than the rather poor vacuums we can create here on Earth. It's worth explaining this a little more.

How you might suck all particles and energy out of a patch of space is an interesting question because it's not just a case of drawing material towards an object such as a black hole, using gravity. It's actually quite a difficult thing to achieve.

> Light is so energy rich that it wouldn't be captured by gravity unless it was passing very close to an object as powerful as a black hole – which we imagine to be a region rich in material and energy anyway.

> So in trying to create a perfect void, the best we can realistically hope for is that the path of light can be bent away from a particular region. Yet there will be light coming towards any place from all directions, so the diversions would have to be present in all directions, handling all other radiation as well.

> At the same time, the diversionary forces, (whether gravity or the 'lens effects supposed of Dark Matter), would need to be perfectly balanced, (ie. gravity has a near infinite range, so unless those forces were perfectly balanced a patch of space that was being cleared by one gravitational source might be occupied by material drawn towards another object if the paths crossed).

Those forces might also have to be quite powerful, which would imply that if the balance was lost, the rush of material into this enclosed void could be dramatic.

Is this type of effect powerful enough to generate a Big Bang? We will consider the possibilities later.

3 –   The Potential for Self-Balancing Mechanisms and Frameworks to influence Spread-Out Material – within the Laws of Physics

The final set of concepts which operate within the Laws of Physics that I want to introduce at this stage, provide *structural* ways in which spread-out material might generate an alternate form of Big Bang event.

Self Balancing Mechanisms

You may recall that there are a small number of factors in existence whose strength settings determine the way that everything else seems to work. Those strength settings could be set to any level within a wide range, so there could be a myriad of different combinations that would each represent a separate type of reality.

Yet from what we can see our Universe only applies one set of these configurations, leaving us with only one flavour of reality, (one set of Physical Laws).

The concept behind self balancing mechanisms is to suggest that there are ways to standardise the configuration of Matter/Energy where more than one type of existence could emerge *at the same time* within a region of space - whether that is a single Universe, or across an infinite existence.

Self-balancing mechanisms fall into two main categories. The first essentially argues that a myriad of different realities may either compete for survival, or find ways to isolate themselves, enabling a single standard to emerge within each zone.

In practical terms, 'Competing for survival' may represent the action of 'chemical/ nuclear' reactions which either destroy the competing elements, or force them to change & adapt until a common standard is achieved. One of the major problems with this type of mechanism is that the enormous distances involved would prevent the different types of reality from interacting. For this reason a number of people will suggest that, contrary to what we think is happening, each of the distant galaxies represents a unique flavour of reality.

The second type of self-balancing mechanism imagines that a chain/circle of reality is necessary to maintain existence. This concept is explored more fully in 10.2.3 below. If any link in that chain either diminishes or 'evaporates' it would have to be repaired or replaced otherwise the whole of existence might disappear. The mechanism would therefore have to provide a way to maintain or replace each section of the chain and in broad terms that would seem to require the transfer of large amounts of material into a particular location – probably across vast distances.

Although this type of notion is more speculative it is primarily this 2nd approach which seems to offer a way of generating a Big Bang event with the type of movement we see in the Universe today.

Yet a 'circle of reality' would imply that a particular form of existence had been put in place without any way of it being able to evolve, because the self-balancing mechanism would effectively be structured to preserve its own existence in that form. In short it would make us dependent on luck or design to put that structure in place.

Of course there's no evidence for either form of self-balancing mechanism, but a lack of evidence is a factor within all explanations of origin, so this is the norm.

Implementing change through a Framework for Existence

Although the original concept of a Framework for Existence was to standardise existence by applying one form of reality everywhere, it is clear that any change to the rules which such a Framework enforced would lead to wholesale transformation.

Of course, this implies that the structure of the Framework, (whether internal or external – per chapter 8.1.3), already exists everywhere. Although we lack direct evidence for such things there are facts that we have uncovered which do support some aspects of these concepts... and of course, a lack of direct evidence features in almost all other mechanisms of origin.

An 'External' Framework would effectively surround everything in existence in order to impose a configuration, which means that each part of the Framework would need to be aware of the settings, in order to apply them. By implication, any change to the configuration settings would have to be communicated everywhere across the entire Universe in order for them to be enforced/implemented.

Conversely, an 'Internal' Framework would allow each part of existence to change itself based on information/instructions received - in a similar manner to the way that messenger particles may communicate the effects of fields to objects which are passing through them. The self-configuration mechanism would have to be built into every part of existence, and once again, any new settings would have to be communicated across the entire Universe.

The intricacy of those communications might vary quite a bit, (and this will become significant later). To illustrate the differences we might say that each setting could be described in precise detail, **or** the instructions could be much more summarised, for instance, referring to 'type a' vs 'type b' if the differences could be derived.

As an example of a summary instruction in its most simple form, we might think of a communication which is a straight choice such as 'Yes or No'. The question behind it might be something along the lines of "Does existence consist entirely of Nothingness – ie. the zero form of something?". If the answer is 'No', it might prompt a split into matter and antimatter everywhere at once.

Once again, the evidence from the Gisin experiments points to a capability for instant communications within nature, which would allow those instructions to be cascaded across the entire Universe instantly.

Yet such a mechanism would prompt deeper questions along the lines of – "What put it in place in that format?" and if new configurations of existence were being communicated - "What was generating those new settings?".

4 - Ways to break the Laws of Physics and their effects

If there is a need to break the Laws of Physics as part of an explanation of origin how might this be done?

Such needs are demonstrated in concepts like 'Universe Inflation', but also in terms of explaining how fundamental change might arise to allow nature to explore every permutation of reality. Within this there are real dilemmas for Materialist thinking.

Physical Matter/Energy is always seen to operate *within* the Laws of Physics: it doesn't change the rules. Yet in order to have a different configuration of reality with different Laws of Physics the rules **would** have to be changed. There is a potential conflict of principles, which goes even deeper when we think that *the rules of an eternal cycle cannot change if it is to remain eternal.*

Although the principle of cause & effect requires precise mechanisms to be in place, a number of Materialist theories skate-over descriptions of these change processes.

For instance the Bang–Crunch theory assumes that the extraordinary circumstances of a singularity would allow new configurations of existence to happen by reducing physical substance to its most crude form – but it doesn't say **how** change would arise.

As we have seen, critics would argue that a singularity would need to remain within the existing Laws of Physics in order to preserve an eternal cycle, so there is no reason why the same material under the same rules should produce a different result. As things stand, there is no answer to this.

The same issue arises with the 'curtains of energy' idea, because the same basic circumstances are generated at the start of each cycle yet they have to produce different outcomes. If randomness & spontaneity are not permitted, this is a problem.

On this reasoning people have to work harder to determine **why** a fundamental change in the rules of existence might arise. Even if we assume that such changes are possible, we don't know which factors could be influential - although we can guess that they might be something extraordinary.

As a pointer to one possibility, Quantum Mechanics provides evidence that different rules can apply to different levels of existence. Could it be that the Laws of Physics might only change at the higher levels of reality which we occupy, and that subtle combinations in the stable lower forms of existence might account for the changes?

It's possible, but if this were true we would also need to explain why the Laws of Physics do not change constantly. Again - it remains a significant gap in the argument.

This line of thinking has led people to consider whether there are **ex**ternal influences, including other types of underlying stuff... (the subject of Chapter 11).

Before looking at some of these points it's useful to consider what might happen once a re-configuration of Matter/Energy was initiated. (There are 3 suggested ways to do this below).

If the structure of any reality is based on certain 'configuration settings', then a change to those settings (eg. through a Framework for Existence), would probably cause the earlier reality to **disintegrate** before the underlying stuff of existence adopted the new configuration – and the associated **new** Laws of Physics.

You may visualise this disintegration as being a gentle 'everything turning to dust' type of effect, but many suppose that the outcome would be much more dramatic. For a brief moment that underlying substance might be freed from many, if not all Laws of Physics, and therefore able to exceed normal capabilities – for instance, potentially allowing this material to go faster than the speed of light.

Suggestions of Universe Inflation probe these scenarios.

Yet the removal of a set of physical laws might be achieved in other ways. If there were multiple Universes across space, sitting side by side, each one may have **already** adopted a different configuration of existence and therefore different Laws of Physics. An interplay or crossover from one of these types of existence to another is very likely to result in an explosive re-configuration of the matter involved, (perhaps even more dramatic than a singularity because a singularity would have already reduced its material to a basic state – a partial conversion).

The final way to change the configuration of reality, touches on something we encountered in Ch. 7 – that existence may occupy different dimensions[40].

M-Theory[22] suggests that there may be up to 7 hidden dimensions plus Time, (11 in total). If physical substance can exist entirely within **our** 3 dimensions it may also be possible for it to exist entirely within some of the other hidden dimensions too, and therefore be invisible to us. A different set of dimensions also directly implies a different set of physical laws.

A cross-over of material from one set of dimensions to another would again potentially generate a fundamental and very powerful re-shaping of existence that would momentarily free the affected stuff from all Laws of Physics.

Each of these 3 ways provides a potentially strong reason why existing Matter/Energy might be re-configured, as well as an explanation of how 'Universe Inflation' might arise to explain the current size of the Universe if the Big Bang represented a single explosive event. Now although the logic is reasonable we should remember that as things stand we have no basis on which to say that any of this is actually real.

This completes the key concepts/principles which we need to bear in mind when assessing some of the specific mechanisms of origin which we are about to encounter. In becoming familiar with them you've done the hard work – well done!

The mechanisms presented in the coming sections all represent alternate ways to explain what happened at the Big Bang. They're fun and they may even be right!

The first group can be categorized as 'elaborate versions' of effects that we know to exist in nature', even if we have never observed them on the grand scale, (10.2.1 & 2).

The second group, in section 10.2.3, are more speculative ideas which seek to exploit the potential effects of other universes or other dimensions – neither of which are proven. However all of these will remain within the principles of cause & effect.

10.2.1   Familiar Effects within a Single Universe and 3 Dimensions

While the previous section outlined some principles/mechanisms that could be used to generate an alternate Big Bang event we need some more specific examples of how such things might be deployed – and this is what we will do here.

The first six scenarios will cause a large burst of matter/energy to suddenly appear at a point in space. The final two will assume that a large amount of 'stuff' already existed in that particular location before it was caused to react/explode in dramatic fashion – leading to the pattern of movement that we observe in our Universe today.

The first of these mechanisms takes us back to an effect that we came across in Chapter 5 - an every-day process in which energy can suddenly materialise in a place where it didn't previously exist. It's the way in which we generate electricity – the scientific principle of electromagnetic induction. Don't be put off by the terminology; the core idea is quite simple!

If you recall, when a magnetic field simply moves near a loop of copper wire, (not touching it), an electric current will just *appear* within the metal circuit of the wire.

In contrast, you can spin a magnet near your hand and you won't feel the magnetic field or any burst of electricity, although the magnet itself may warm up a bit. So we need certain conditions to exist before the Induction effect can take place.

The two principal explanations for induction are that either the energy flows through the invisible magnetic field into the wire, or energy simply emerges from nowhere within the wire to create an electric current where none existed before... we don't really know which. (Most scientists will generally prefer the former).

The mere presence of the wire in the vicinity of the magnet **creates the right circumstances** that enable the electric current to appear.

Yet we don't believe that enormous magnets & coils of copper wire were set up just before the Big Bang at the beginning of our universe, so the principle would have to point to deeper capabilities within existence. For instance...

As the metal in the copper wire is made of atoms/sub-atomic particles which have their own small fields of influence, we could say that the generic mechanism of Induction is an interaction between different types of field.

On the scale of a Universe and the Big Bang, enormously powerful fields would have to exist and interact in order to cause a mass of energy to suddenly appear at a point in space via an induction process. The burst of energy would probably arise in the place where the two fields were interacting.

We don't know what might generate those powerful fields, (it might just be the presence of scattered pools of energy), but the key thing is that any process of induction requires *something* to exist in advance.

So conceptually, while a mass of energy may suddenly appear at a particular place as if out of nothing, it would actually represent a **transfer** of energy from somewhere else. Returning to our porridge analogy, the differences can be illustrated like this:-

Why did the pan get hot? We can say that :-

- it was just lying around when it suddenly became hot for no reason, (spontaneous pure creation of heat/energy), or

- God came along, and either ordered it to be hot so it did, or He moved the pan onto an Induction Hob. In neither of these instances would there be any direct input of energy from Him, (pure creation – but not spontaneous),   or

- the pan drifted onto an induction hob where it began to get hot, (natural cause & effect within eternal existence).

Of course, for this final scenario to work, the pan, energy, and hob would have to exist beforehand, and a way would need to be found to bring all of the factors together in order to activate them.

Within the notion of an eternal existence, but without considering pure creation which is covered in the next chapter, the other 5 speculative ways for a huge amount of energy to suddenly appear at a point in space are more straightforward :-

- the generation of an unknown type of **'energy vacuum'** - equivalent to a pressure difference, which causes nature to suddenly fill that void from pools of energy that surround it – but in a fast and dramatic way, or

- there may be a **'rip' in space** that causes a channel to open up between two very distant places, allowing material to flow between them; (Wormhole theory generally envisages a bending of space in order to bring distant parts of the Universe close together).

- the **splitting of nothingness into balanced opposites** (matter & antimatter) which assumes that nothingness is actually the zero form of something, and contains a hidden source of energy/matter. .. but this also needs a trigger.

- the suggestion from advocates of M-Theory that other hidden structures, such as vast **curtains of energy**, (bigger than a Universe), exist beyond our sight.

- the earlier notion of a **'self-correcting'** structure where many parts of existence have to be kept in balance to prevent the whole 'edifice' collapsing

To clarify the option surrounding a rip or channel in space I expect that most people would imagine that empty space would need a structure/fabric which can be torn in order to create a channel. Yet such a fabric is completely unproven, even if there are mathematical models based on Einstein's Theory of Relativity to support the idea.

An alternative would be the equivalent of a sea current that physically transported material along a route – yet it's hard to envisage this working at a sufficient rate to deliver a Big Bang type of effect. Material would potentially have to travel at speeds up to, or even beyond, the speed of light across vast distances without reacting with anything en-route to suddenly create a burst of energy somewhere.

A more effective version of this idea is that a rip or channel may link our reality to other hidden dimensions (10.2.3) meaning that the distances travelled could be virtually 'nil' while at the same time the explosion could be disproportionately powerful as transferred material was re-configured, (as we considered earlier).

To complete our list of candidate mechanisms for the Big Bang within the principles of cause & effect and a single Universe, we need to add two other concepts which we have already come across. These do not require huge quantities of Matter/Energy to suddenly appear. They simply transform a mass of stuff that already exists.

Firstly, there are **'chemical/nuclear processes'** - my term for all of the other 'normal' reactions that we know Matter/Energy to be capable-of in its various states. This would include supernovae and even exploding singularities, but would obviously exclude the six suggestions above to avoid duplication. All of these would have to operate within the known Laws of Physics within the different layers of existence.

Secondly, we can consider **altering the configuration settings** of existence using an internal or external Framework as discussed earlier. The key to this idea is to explain how new settings could arise without Thought/design.

While the 'normal' purpose of a Framework is to standardise existence on a particular format, the suggestion that new settings could be fed into such a mechanism implies that it might also be a driver for wholesale change through information being translated into physical effects.

You may find it difficult to believe that natural processes can work on the basis of information received but the fact that our bodies can sense and react to what is happening in the world around us using electrical signals is proof that control mechanisms using coded signals can arise naturally. If you are a Materialist then this must somehow arise from matter/energy alone, and there is also the scientific concept of messenger particles to add support to this notion.

Of course, if you are a Dualist then you are likely to attribute new settings to the action of a different type of stuff – Thought, but this will be covered in the next chapter.

In the context of this chapter a Framework mechanism would therefore be dependent on new settings arising without Thought, and because we do not see nature 'flipping its configuration' on a regular basis we have to assume that such a change would be the result of unusual circumstances. What could these be? Well, we can only point to some generic possibilities:-

1. The Framework may not only look for parts of reality that are threatening **not** to conform, (in order to reinforce the standard), but may also look for new settings that seem to work in a better way... which implies a degree of awareness or at least a sense of direction given that there **isn't** constant change.

2. The Framework may occasionally be directly affected by its environment, causing a change to its settings for better or for worse.

3. That the Framework is linked to a place which generates new settings.

To some people option 3 would represent something similar to a control centre for the Universe which would go against the principle that all parts of nature look after themselves... and for many this idea will also smack too much of God. However that doesn't mean it can't be right.

Option 1 seems to require the Framework to have a sense of awareness or direction which we see in the evolution of life, but this adds to the requirements for a Framework making it harder to achieve. Option 2 is just worrying ☺ - but again there is no evidence to demonstrate that this is real or happening, so we can relax!

This completes our generic list of suggestions, excluding multiverse or multi-dimension concepts. For the purposes of our onward exercise, we need to assess whether each of these mechanisms might be able produce the expected outcome before gauging how likely each one might be to exist, given the evidence available. This will be summarised in Chapter 12, but for the moment we can apply a greater level of critique to these basic ideas.

10.2.2   Deeper Challenges to some of the Key Options

The previous section gave us a basic list of 8 mechanisms which might explain the origin of our Universe within the 3 dimensions we occupy.

All of them expand on well known features of our reality, but they push those concepts to the limit, making it highly questionable whether those things might actually exist on the grand scale. Yet because they all suffer from a lack of evidence to prove their existence, our assessment of viability needs to determine whether there are other problems with these ideas.

Can We Realistically Expect Any Cycle to be Eternal?

As a background requirement, all mechanisms that are part of an eternal cycle must guarantee to maintain the cycle every time it is run. Unless the cycle is robust, it would break down and fail to meet its conceptual purpose, making it dependent on luck to achieve our outcome. This doesn't just apply to the mechanism. If the process is dependent on a particular start point, we also have to explain how existence keeps on returning to that that starting position without luck.

People tend to 'gravitate' towards the simplest solutions – assuming they are the best, however simplicity is only good if it does the job.

A single explosion arising from a singularity as part of the Bang-Crunch cycle is a lot simpler than a myriad of co-ordinated explosions, but as we have seen, its assumption that a singularity would always produce a huge explosion together with a new version of reality without ever breaking the cycle, is questioned on many fronts.

It is not clear that chemical/nuclear explosions would be able to generate a blast on the scale of a Big Bang. The largest explosions that we see today are supernovae and these do not overwhelm their own galaxies let alone a universe. If we also ramp-up the scale of object to those which most closely resemble a singularity, (Supermassive Black Holes[36]), we do not seem to have any greater success.

In short, it seems that they <u>don't</u> explode. Black Holes are now thought to vent their surplus energy[37] (either as a low level 'glow' or sometimes as beams of radiation from their north and south poles when they are 'consuming' suns and planets). So if a Singularity resembled a Black Hole it **wouldn't** be guaranteed to trigger a Big Bang.

It might be argued that this venting is because black holes have a drip-feed of new material which they consume. Stars and planets circle for a while before falling-in, and they come one at a time not in huge numbers. It is therefore difficult to see how the 'gathering stage for a singularity would be any different, and therefore it is unlikely that their 'digestive capacity' would be overwhelmed.

There are an estimated 200 billion black holes and they won't all arrive at the same time. Even if they did, there is no explanation why they would produce a new version of reality or how the level of change should **never** be enough to break the Bang-Crunch cycle – especially when we see that it **has** been broken.

The curtains of energy idea doesn't suffer in these ways and provides a reasonable process both in terms of generating a start point and providing a mechanism with a degree of reliability. Yet it is entirely dependent on a made-up structure for which there is absolutely no evidence, although it illustrates that viable concepts are possible.

So does the idea of multiple explosions on scattered material fare any better?

Let me answer by saying that the myriad explosions idea is **not** the basis of a repeating cycle. They primarily explain the distribution and pattern of movement that we see in the Universe. As a concept, it therefore has to be **allied to another** mechanism which might generate fundamental change on a regular basis – perhaps using something like a Framework for Existence with new parameters.

Hopefully this illustrates that there will be numerous perspectives to cover when we come to assess each scenario, and any of these 'angles' might present new challenges.

The Breaking Point for Credibility

Where does credibility become seriously challenged? That is obviously a very personal judgement, and one that will vary with each scenario, but we can probably identify a few generic factors.

As mentioned several times already, explanations of origin based on an exploding singularity previously dominated many peoples' ideas because they were the only ones which had substantial evidence to back them up – their key advantage. Yet we have seen that recent scientific findings have undermined them in several ways, leaving all theories of origin, (religious and scientific), with some key aspects that lack any evidence. This is no longer a distinguishing factor.

The **number** of such problems within each concept may therefore be a better pointer towards credibility, but to make that call we initially have to assess all mechanisms on the basis that they might be real.

Beyond this, we may have greater confidence in mechanisms which are evident in nature rather than entirely made-up. A process of Induction may therefore be more believable than a Framework for Existence, or hidden 'Curtains of Energy' .

Yet we also have to consider the likelihood of whether each mechanism is capable of producing our outcome – and some are clearly much stronger than others.

A more subtle judgement is whether each scenario makes sense – not in terms of mechanics but in terms of overall context. Perhaps I need to illustrate what I mean.

If we suggest that a channel in space transfers huge amounts of Matter/Energy to a particular location before causing a Big Bang, what's the purpose in that transfer?

Why didn't the event happen where the material originally lay? There may of course be different answers to this, (eg. to concentrate the energy so that it would explode in the necessary way), but a reason does need to be provided which may in turn expose other needs - in this case the need for several channels not just one. These cascading implications have to be considered.

As another type of example, we might say that if the concept of 'splitting nothingness' requires huge amounts of energy to make it happen, but it produces little extra matter or consumes more energy than it generates – why wouldn't the original stock of energy be used to create a Universe directly? Again, it might be argued that the process needed a catalyst of some sort to make a Big Bang happen and that might only come from the splitting of nothingness – but such explanations add layers of complexity that can challenge credibility – especially if there is no evidence.

It is these types of judgement that we need to make about any scenario. Yet despite particular failings we should consider the principle within each scenario - which may be most useful factor, especially if it can be used in combination with other ideas.

10.2.3   Scenarios for a Multiverse and/or Additional Dimensions

I have separated these notions as they seem to enter an entirely different level of speculation, by drawing together ideas about multiple universes (within our 3 dimensions); parallel universes (in different dimensions of existence); and suggestions that a universe can extend powerful influences across huge distances in space.

The concept of multiple universes, which effectively sit side by side within our 3 dimensions, comes from two ideas.

The first is that eternal existence could simply be without limit and that other Big Bangs could occur elsewhere too – creating an infinite number of other universes with different configurations, (ie. other types of reality with different Laws of Physics).

The second imagines that the process of very rapid *inflation* which is supposed to have occurred immediately after the Big Bang, didn't stop, thereby creating a universe that is vastly bigger than the one we imagine today. Yet this 2nd scenario still starts with a **single** explosion at a point of prime origin, which has a number of implications:

1. Everything would be part of the same Universe, and the material that exists beyond the visible galaxies is less likely to have a different configuration and the Universe would be very unlikely to collapse again. There would be no obvious cycle with which to explore other forms of reality so we would be dependent on luck to have achieved our version of existence.

2. As the material in the Universe would continue to spread out in all directions becoming ever more thinly spread, (ie. more distant from their nearest neighbours), it becomes difficult to rationalise how a myriad of new Universes could form from that distant material.

3. If you imagine our Universe as an expanding ball, and then imagine other universes of equal size forming on the outer edge of that circle, and then imagine how many of those other balls could fit around our Universe, you will be able to see that the amount of material needing to emerge from the Big Bang would need to be hundreds of times larger than all the material we perceive today, (including atoms, dark matter, dark energy, background radiation etc), and yet that would still be a very limited number of Universes with which to explore the different permutations of reality.

In terms of avoiding the need for luck in generating our outcome, an eternal & infinite Universe is a much stronger concept for nature to explore all permutations of reality. So let's pursue this.

If you are not persuaded that a singularity is guaranteed to generate a new version of reality, the greatest potential for creating a new Big Bang explosion would be where different configurations collide and produce violent reactions. This would occur on the fringes of each Universe, but the implication is that the volume of material would be limited at these touch points, and if Universes were that close, such reactions would be occurring all the time – yet there is no evidence for this.

However alternate versions of this scenario, mentioned earlier, take us into the realm of Parallel Universes by suggesting that large amounts of physical substance are contained in hidden dimensions of existence which occupy the same space as our 3 dimensions. On this basis, a rip or channel across dimensions might be rare and might also have enough material to generate an entire new Universe.

Let's explore these notions in two ways: firstly as an extension of the singularity and 'curtains of energy ideas', and secondly in the context of a 'circle of existence', as ways to trigger a Big Bang.

Extending the Singularity and 'Curtains of Energy' concepts

There are three problematic issues at the heart of both of these ideas, which might be resolved by the prospect of parallel universes, (ie. vast quantities of material being held in other dimensions which we cannot detect) :

1) How can the size of the Universe can be explained if it exploded from a small point in space, (ie. how can material break the laws of physics to break the speed of light many times over)?

2) What could cause the material to adopt different configurations/versions of reality with different laws of physics to guarantee our version of reality?

3) What could provide a sustained level of new energy to power the accelerating expansion of the Universe over the past 5 billion years and with en even effect everywhere?

If real substance is held in other dimensions and it finds a way to cross into our dimensions, then it will not only provide a source of material for the Big Bang but will also have to adopt a fundamentally different configuration if it is to survive in our 3D space. In undertaking that change we have already speculated that the material would need to be broken down into its most fundamental state before being re-configured in an environment with different laws of existence - ours.

In one sense, a fundamental re-configuration of existence might explain why a singularity exploded in a very powerful way instead of venting its energy like a Supermassive Black Hole. In that moment of transition it might be free of all Laws of Physics and therefore able to exceed the speed of light.

An alternative way to explain the size of the Universe would be if the transfer occurred across a very broad area, but still powerful enough to make it travel outwards.

If a transfer like this occurs once it may also happen again, perhaps in a less powerful form to simply drive the accelerating expansion of the Universe.

As we don't see powerful channels of new energy stretching across the Universe from a point in space, we might envisage a more gentle flow that also permeates into our space across a very broad region. The more gentle flow might be explained because the original transfer occurred in a void, whereas now there is some material, so the 'relative pressures' might be less.

We therefore have to ask how a channel could suddenly open between different dimensions when the structure of existence would seem to be so stable?

As I have just indicated, an initial suggestion could be that the presence of a void might cause a significantly powerful pressure difference.

Alternatively, the power of a singularity might be another way to achieve such a breakthrough. In this scenario, when a 'Big Crunch' happened it might also rip a hole between dimensions and transfer the whole singularity to the other dimensions – generating a Big Bang explosion there, resulting in a new configuration of reality and a period of Inflation. The next Big Crunch might transfer the re-formed singularity back into the original dimensions – representing an extended version of the Bang-Crunch cycle.

However if a hole between dimensions is dependent on a void or a singularity there isn't a way to explain how another powerful and extended burst of energy could arise from other dimensions part way through the cycle, to accelerate the expansion of the Universe 8.77 billion years after the Big Bang, (thereby ending the cycle as well).

In contrast, the 'curtains of energy' idea is not affected in the same way, because those curtains would now be envisaged as vast pools of energy within discreet dimensions, separated by physical material in our 3 dimensions.

If material in our space became too thinly spread, then the other dimensions might come together to generate a new burst of energy in our space, pushing the other dimensions apart again... conceptually. This I believe is what has been proposed in the Epkyrotic Model[40].

This may again account for an Inflationary effect, but equally, if the clash of these gigantic curtains occurred across a vast area, (say 75% of the volume of our Universe), then no inflationary effect would be necessary.

It may also be envisaged that a thinning Universe might, under certain circumstances trigger a partial flow of energy into our dimensions again, to power the acceleration over the same broad area of space. It would also remain true that in this scenario the acceleration effect wouldn't break the eternal cycle of experimentation with new configurations of reality.

Circles of Existence

Let's imagine a series of ball-shaped universes arranged in a circle - each one sitting between 2 others, side by side. For simplicity let's also assume that they have the same configuration of reality.

A circle tends to imply that these balls will influence each other in order to maintain the ring shape, and that Dark Energy might be part of the mechanism which holds them in position. (Other shapes may be envisaged but a circle is useful to illustrate the point).

So let's consider how structures like this could produce a Big Bang.

If the different universes were solid we could imagine them bouncing off their neighbours within the circle, but the gaps between galaxies and solar systems are so vast that the outer regions may simply merge together. On the opposite side of the Universe to an overlap we might find a gap, so the potential influences may be the equivalent of a static electricity discharge in the overlap zone, or effects related to an energy vacuum on the other side.

If we were to say that different universes also had different configurations of existence then any interaction would be likely to be much more violent.

In terms of ordinary thinking, all of these effects might take place within our 3 dimensions, but the suggestion from M-Theory[22,38] is that there may be as many as 10 physical dimensions plus time, which opens-up new levels of speculation.

In short, the circle of reality may cross dimensions and as we don't see instability across the universe, this may suggest that the different forms of existence are part of a stable structure where the different elements form a chain of mutually dependent existence. The idea may sound bizarre and unfathomable at first hearing but it's actually quite straightforward. The best way to illustrate this is with an example.

When we look at the similarity of how planets and asteroids are arranged in orbits around a sun, it's easy to draw a parallel with the layered structure of atoms where electrons 'orbit' a nucleus. We see the same pattern in the way that stars are seen to orbit around supermassive black holes[36] at the centre of a galaxy and so it can seem like a constant theme at all levels of existence.

The notion has therefore arisen that the solar systems or galaxies of our universe form the *atoms* of a larger reality. (This has even been 'demonstrated' in some movies and TV ads).

The solar systems and galaxies in that larger reality may, in their turn, form the molecules in an even bigger reality and so on - effectively forming a chain of reality that was an unending line of existence, disappearing into infinity... unless it was arranged in a circle.

However to form a circle we would have to suppose that ultimately the galaxies at the top of that chain bent around to form the atoms in our reality, **forming a circle of existence which you can never get to the bottom of**.

Within the 3 dimensions that we occupy it would seem impossible that a galaxy in a **bigger** reality than ours could form an atom in our finger, so the only way I have been able to make this concept work, even vaguely, is to suggest that each layer of this chain of existence has to occupy a different 'dimension' of the same space.

Put another way, it may be possible for something very large in one dimension to be very small in another. If you run with this notion of 'a circle of reality', then *every significant part of every Universe would form a necessary part of the chain of existence and if any part disappeared the whole edifice could collapse and disappear*.

Well, yes, I know... but the whole notion is fanciful so please forgive my attempt to rationalise it! So how do we link this to the notion of a Big Bang?

In the 'origin scenario' within our 3 dimensions, if one universe in the circle disintegrated, evaporated, or dissipated to near nothingness, the circle might contract – becoming smaller to plug the gap, but across dimensions any break in the chain is likely to threaten the **whole** of existence, because each link in the chain would literally form part of the structure of everything else.

A self-balancing mechanism would be needed to repair or replace weak sections of the chain and this, it is suggested, might lead to the generation of a new Universe perhaps through some process similar to Induction. However there are problems with this idea.

Firstly, if any part of the chain needed to be replaced it would have to be done with the same type of material that occupied the same dimensions as before, in the same location and probably the same configuration settings. In other words the chain of existence would be a fixed format and therefore each link in the chain would have a prescribed configuration. On this basis there would seem to be no opportunity to experiment with different settings and therefore the scenario is likely to require luck in order to incorporate our specific form of Universe as part of the overall 'design'.

Secondly, if the purpose of a self-balancing mechanism is to prevent the disintegration of the chain of existence, it would seem very strange that it would permit the situation to become so severe as to warrant the formation of an entire new universe. The level of instability should never become so severe as to warrant a Big Bang, however we have no idea how sophisticated such a mechanism might be, if it exists at all.

So this notion of a circle of reality offers some potential, especially if it crosses dimensions, however it is equally possible that other dimensions may exist but there needn't be a mutually dependent circle of existence. Stuff may simply occupy different dimensions without any pattern or dependent structure.

This is because there's **no** evidence for other universes or even other dimensions as things stand, but there is nothing to disprove them either, (especially as we can't see the edges of our universe). We have no basis for saying what might happen other than in the broadest of terms.

From this you will hopefully see that circles of reality, multiple universes, and parallel universes all add to the list of 8 options we saw earlier to complete our list of options based on cause and effect, and if new ones arise, (as they probably will), we can try to see where they might fit into the overall scheme of things.

The next chapter will consider how these options could be influenced/improved if there were other types of stuff that enabled spontaneity & randomness to occur. This will also include the possibility of spontaneous pure creation, as well as showing how different forms of Thought might influence events.

As a slight blurring of the concepts we will also cover the potential for physical thinking beings to influence Big Bang events, as we know that physical beings are a reality which Materialism cannot deny.

This is far from a silly idea if we accept that existence has been dynamic, eternal and subject to a succession of Big Bangs which can try out different permutations of existence. In these circumstance it is entirely possible that an intelligent species might arise in an earlier Universe to influence the next Big Bang event - ours.

However for the most part the influence of Thought will be a metaphor to consider the potential which other types of stuff may offer – the underlying capabilities of spontaneity and randomness.

We should also recognise that such stuff may arise in a variety of forms in much the same way that Matter/Energy can take different forms – some more sophisticated than others.

11 Scenarios for Origin (incl. Spontaneity & Thought)

This chapter looks at scenarios for prime origin which go beyond the expected capabilities of Matter/Energy, to include spontaneity, Thought, and 'Pure Creation'.

This is not 'pie in the sky' stuff.

As a start, thinking beings are a **fact**. *You* are a testimony to that, so ***Physical*** Thought (based in Matter/Energy), is real and not disputed. There are also a number of reasonable scenarios where physical beings, similar to us, could have influenced formative events such as the Big Bang, and I will run through this thinking in sec 11.2

However if existence is underpinned by a 2nd type of underlying stuff which brings additional properties with it, there will be a number of other ways in which the process of origin may have been influenced. This other type of stuff is likely to take a variety of forms, which may generate 2 other types of Thought, and each of these may provide a distinct means of influencing physical events, with different levels of sophistication.

To illustrate the point, Matter/Energy can present itself to us as:

- a raw state — energy/heat;
- something basic but tangible — atoms that are 'solids, liquids, & gasses';
- something sophisticated — a living body or a constructed ship.

So it's not unreasonable to suppose that a second type of stuff may do the same:

a) A 'raw' or basic version which just enables spontaneity to get new things going: either as a source of new existence, or a trigger for new activity.

b) A 'medium-level' version providing a basic sense of awareness, or direction.

c) A 'high level' version which generates full-blown Thought.

As there seem to be a growing number of scientific theories which now **require** spontaneity/randomness to keep them alive, the ability to provide just a basic level of capability at the Big Bang, (rather than sophisticated Thought), may be sufficient to bridge any early gap, while potentially allowing this other stuff time to evolve as well.

We see such needs in theories which use 'chance' to explain changes in our Universe, (eg. new configurations of existence), and also in more specific issues, such as the ending of an eternal sequence like Bang-Crunch.

> Remember - 'chance' isn't a feature of 'cause & effect' so it cannot arise from Matter/Energy. As a result, materialist explanations of Evolution, or other such theories, should either be re-written as an inevitable process with one outcome, or they need to include some other type of stuff to make them work.

Pure Creation is another serious possibility considered by scientists as well as the religious because it is one of only two core options for origin, (the other being eternal existence). Yet many people do **not** believe this to be a capability of Matter/Energy, so Pure Creation is also linked quite heavily with notions of a second type of stuff.

I will look at the science-leaning perspectives of Pure Creation in 11.3 and consider the basic religious concepts in 11.4.

Concepts based on spontaneity, Thought, or Pure Creation will appeal to some people and will be anathema to others but in an age dominated by science they do undoubtedly challenge established thinking, and the immediate retort will be to point out the 'lack of proof' for the key mechanisms involved, (including God)... until it is realised that **all theories of origin fail on this point**. As we have seen, from a neutral point of view, lack of evidence is not a distinguishing factor between theories of origin.

Indeed, when you set all of these uncertain mechanisms side by side and honestly assess their ability to overcome the challenges of origin, people will often find that some ideas which they previously dismissed start to look more credible, while other ideas which they originally favoured could start to look less convincing.

We should remind ourselves that the reason why concepts about a second type of stuff persist, is that many people feel that Matter/Energy cannot explain origin by itself, primarily in relation to 6 main issues. The first 4 are fairly self explanatory :-

- Providing **a source of true change** – particularly to the configuration of existence.
- The need to **avoid luck** in providing a solution to our origin.
- Explaining how sterile chemicals can apply physical **control** in complex repeating processes, varying their reactions/outcomes in relation to changing circumstances, (incl. the workings of living cells grown in a test tube without a brain - Book 2).
- The need for **standardisation** across the universe to ensure that things operate in the same way everywhere.

The other two issues require a bit more explanation.

Number 5 boils down to scientific findings from recent decades which increasingly suggest that nature may need **a basic sense of awareness or direction**. Sterile chemicals shouldn't need this yet, as examples, we can point to

- particles in the double slit experiments appearing to know whether one or two slots are open, reacting accordingly, and potentially aligning their results *after* the experiment has occurred.
- experimental results demonstrating faster than light communications and a sense of mutual awareness between paired particles.
- Evolution – which shows a pattern of general improvement and increasing sophistication – rather than degradation and decline as might be expected from the 2nd Law of Thermodynamics, (see Book 2).

Finally, as we've seen in the last chapter, many scenarios for origin require something to push the natural environment beyond its normal level of performance so as 'to **tip the balance**', perhaps by allowing natural physical mechanisms to operate more effectively, or on a scale that Matter/Energy wouldn't normally achieve, (for instance, the interaction of fields to generate induction on a Universal scale).

If you feel that these issues are real and that they require something like spontaneity, randomness, Thought/design, or Pure Creation to overcome them, then you must either

a) rule out those theories which require them, or

b) accept that the hidden factor may be real, even if we can't detect or explain it.

As all theories have issues (the majority relating to the 6 items above), a hard-nosed approach on option 'a' would leave no theory available, so people generally find that they are prepared to consider some of the unlikely possibilities.

Having given the overview, I want to briefly return to one of the points above – the possible need for nature to have a sense of awareness or direction.

As the latest scientific findings into the workings of Matter/Energy suggest special properties for some sub-atomic particles which may even be breaking the known Laws of Physics, we might expect those strange effects to appear wherever those particles are present - but that isn't the case.

Neither can we say that these strange effects should be confined to the bubble of an atom because many of these sub-atomic particles 'roam free' in the wider world. They are everywhere, and yet we don't see their effects at our level of existence. Reality tells us that these capabilities can't simply follow the different types of particle.

Quantum Mechanics has been trying to explain this, but it is a relatively new discipline which is still finding out lots of new things, so it can only offer speculation, not any proof, and some of this speculation is quite broad-brush and unspecific.

For instance, rather than saying that things 'know their place' there are suggestions that particles may effectively be following the 'path of least resistance' yet there isn't really anything to cause a 'resistance' as such. (The relevance of this will become apparent below)

As an alternative there are suggestions that where there are multiple possible outcomes, (perhaps even a near infinite number of options), all possibilities might be explored but most will 'cancel each other out' leaving only the outcomes which we would normally expect. (In broad terms we can refer to this as Feynman Theory).

To illustrate such thinking, (even though there are many problems with the scenario), let's imagine a particle travelling from 'a' to 'b'. While **we** might realise that the fastest route is a straight line between them, there is no reason why an unthinking particle should be aware of this or why it should follow the most direct route. It may therefore follow a curved path, or one that zig-zags,, doubles-back, or goes in any number of complex and circuitous ways.

This theory suggests that if a move in the right direction is 'positive', and a move in the wrong direction is 'negative', those paths which detour, loop or double back would effectively cancel themselves out, to just leave the particle with the ones that comply with the known rules of existence. In a sense it is a mathematical way to illustrate resistance to one course of action as opposed to another. The trouble is that this 'consensus on a viable route' isn't built-up through a period of experimentation; **the particles just seem to know instantly**.

Although there are many questionable assumptions and other problems with this example, I offer it to show the underlying principle, that somehow a type of resistance can emerge through complexity or other 'detour activities' which might generate a sense of the easiest course of action. Yet all talk of 'sense' in this respect implies a degree of awareness.

More generically we can suggest that nature seems to have followed a path of increasing sophistication, complexity and effectiveness, but lifeless chemicals shouldn't have any need for a sense of direction. They should have no sense of purpose, so why should nature follow a path of continuous improvement?

The deeper you probe on such arguments, the more it can seem that the stability of our real world relates to some measure of awareness – even if that's not full-blown Thought. Does this point us towards a 'middling' form of the second type of stuff?

Hopefully you now see the logic which drives the different perspectives.

11.1.1 Types of Thought and their Characteristics

Speculation that there are different types of stuff which underpin existence has led to 3 main concepts of Thought:-

1) As a minimum capability accepted by materialists, ***Physical Thought*** would be based entirely in Matter/Energy, so it would be bound by the principles of causality, (whether being achieved by electrical signals, vibrations, patterns, etc).

2) Thought that was based on the stuff of randomness & spontaneity, might be considered as '***Intangible Thought***' which <u>wouldn't</u> have to imply God, but wouldn't exclude that possibility either.

3) '***Pluralistic Thought***' would be a mix of the two types of stuff above.

Physical and Pluralistic Thought could only operate if *physical matter* already existed, which means that physical beings such as ourselves would have to be using one or other of these types of Thought, (due to the arguments presented in Chapter 2.3).

Intangible Thought is the only form which would be able to exist without a physical body: either ***alongside*** Matter/Energy, (but separate from it), or in ***complete isolation***, (before any physical matter existed at all).

It's easy to see how Intangible Thought could be equated with some forms of deity however this is not the only possibility; for instance, this could be a model for an intangible thinking being without special powers, or a 'life force', or a soul, rather than a god. All are conceivable. The key factor for me is that Intangible Thought is based on stuff that is *within* the totality of our existence, and which may also help to explain life and some of the mysteries of the Universe.

Anything which is *beyond* our existence is also beyond the remit of this book.

> I say this because some people speculate about another type of Thought, (which we might call 'Spiritual Thought'), that has to be based on something which is outside/beyond our Universe. This concept essentially argues that the special powers attributed to a Creator God require something entirely different: a 3rd type of stuff which is unrelated to the stuff of mortals.
>
> Because this notion effectively separates God from both the physical universe & ourselves, plus there is also no real concept of
> - the additional properties it might have, or
> - how it would interact with anything in our Universe,
>
> then there is no basis for ***any*** speculation about it. I merely mention it here as a vague notion for completeness.

Returning to factors which could demonstrate a thoughtful influence within our physical Universe, all 3 forms of Thought offer ways to explain how Matter/Energy can be made to deviate from its inevitable chemical path.

Physical Thought can do this by saying that although our own thoughts and ideas may be inevitable, they are **following an alternate path to physical matter**, and it is only when we get an interaction between the two that we find that the natural environment changes course. In other words, Physical Thought would be acting as an **ex**ternal influence on physical matter, even if it is made of similar stuff.

This might work if we are also able to explain ***how*** the two things came to be on different paths, (given that they emerged from the same stuff). Yet purists would still say that two inevitable paths which interact can still only lead to one inevitable outcome.

In contrast, Intangible and Pluralistic Thought offer true spontaneity and randomness in order to facilitate genuine change, (ie. a way to start something new rather than simply representing the next step in an inevitable sequence), because they draw additional capabilities from a different type of stuff.

We can extend this thinking to say that because our thoughts are capable of 'analysis', 'design', and 'control' they are **pro-active**, and not just *reactive* as cause & effect would imply. Put another way, the process of analysis **isn't** one of reacting to a cause.

> Observation might record that 'a' followed 'b' but in trying to understand *why* that happened we have to be creative and invent principles & concepts.

Hopefully the above points illustrate the different types of Thought that may be available and the limitations which each would have to accommodate, however they could only influence particular events if they were present at the right time.

Before exploring the possibilities for origin in 11.2 I want finish setting the scene for some of the other ideas covered by this chapter.

11.1.2 <u>Some of the Characteristics of Pure Creation</u>

I'll begin by re-stating that I define *spontaneity* as being 'without a cause', and I define '*pure creation*' as emerging from absolutely nothing.

These are separate concepts, and although they can work together, they don't have to.

If pure creation is possible at all, it comes from nothing, but it may or may not have a cause. If it doesn't have a cause, (not even God), then it would be truly spontaneous, but the traditional concepts of a *Creator God* suggest that a cause should be possible.

Given that pure creation means 'coming from absolutely nothing', a *cause* would simply mean 'generating the right the right circumstances for pure creation to occur', but nothing that already exists could directly contribute to the material being created. This is because pure creation adds to the overall pot of existence.

Pure creation is therefore a very distinct type of event, even for God, so

> any stuff that came from, (out of), God must have already existed, and any comments that 'God created something out of'... would actually refer to transformation not pure creation.

Of course atheism will seek a version of this concept that doesn't require a deity, and in this respect it's conceivable that the circumstances for pure creation might just happen naturally in a dynamic pre-existing environment, and not require God.

Yet if it was just a case of waiting for the 'right circumstances' to arise, we might expect to see many more instances of pure creation occurring in the 13.77 billion years since the Big Bang event... yet there is no evidence of this on a major scale.

Put another way, if we liken the Big Bang to firework rockets going off, there are **no** overlapping circles of stars as if two went off close to each other. Materialist explanations have to accommodate this and we will look at some arguments in 11.3.

However, talk of 'circumstances naturally arising' requires something to exist already, and some scenarios for origin *don't* include a prior existence. In this case pure creation would have to be ***spontaneous*** – yet this presents us with a number of conceptual difficulties which we will also explore in 11.3.

For the moment let me just say that the case for God **hasn't** been removed by Materialist thinking or scientific evidence. God clearly represents more than a set of gaps to be plugged, but such gaps are a way to indicate that an explanation is incomplete and that other things may be necessary.

Spontaneous Pure Creation presents us with 3 seriously challenging issues :
  a) The need for a capability to be truly spontaneous instead of a trigger,
  b) The ability to generate something out of absolutely nothing.
  c) The need to avoid luck in generating our specific outcome.

Neither spontaneity nor pure creation (a & b) are generally associated with Matter/Energy, which is why there's considerable scepticism about such claims. Matter/Energy always requires a cause, and always has to 'balance the equation' – because we believe it has never been seen to do anything else.

Yet even if these capabilities were possible, pure creation, (especially in the context of Big Bang), would generally be perceived as a one-off event which would seem to require luck to determine its outcome. Explanations are needed to overcome this.

Few prominent scientists openly suggest that a physical environment may be capable of spontaneous pure creation, and those who do must try to find ways to bridge the gaps. One of these is Stephen Hawking, and we'll consider his statements in 11.3.

The main driver for considering something coming out of absolutely nothing is the logic of origin which makes this one of only two possibilities, however such logic would be enhanced if there were some demonstration that pure creation may be real at all and not just wishful thinking... and there are some real factors which point to this.

The two most familiar bits of evidence are

  • the Big Bang which indicates that our entire Universe may have come from a single point in space (which is in-keeping with our perception of pure creation – even if this particular event might be explained in other ways).

  • that a myriad of new thoughts seem to emerge in our heads and we don't believe that a stock of 'empty thoughts' is being converted into useful ideas. Remember - electricity doesn't think and nor do neurons as far as we can tell. The precise pattern of signals in the brain also indicates control.

While the pure creation of ideas would only generate intangible things, (not Matter/Energy), it remains a potential demonstration of pure creation. The trick would then be to explain how that intangible stuff might provide a source of our physical reality – and there are potential ways to do this which we will consider in 11.4 below.

All I am saying is that a rationale exists to explain both the physical and non-physical aspects of existence, but neither is complete and therefore both may have an equal chance of being true or false, depending on how you interpret the evidence.

If there are barriers to a full explanation by means of pure creation, then we need to identify what those might be, plus the capabilities that would be necessary to overcome them. We would then be able to explore what options might be available.

11.2 How physical beings may have influenced the Big Bang

Explanations of how the Big Bang and our universe 'came to be' are much more likely to be successful if the underlying stuff of Thought and of physical Matter/ Energy can be seen to work together, rather than one trying to dominate the other.

The advantage of Thought to this scenario is that it *could allow nature to target outcomes without exploring every permutation of existence*...

At a crude level, spontaneity & randomness could be the perfect complement to cause and effect, allowing a controlled measure of change & development within a relatively stable environment. Yet wishful thinking doesn't make this true.

When looking for examples we all know that physical beings such as ourselves **can** change the natural course of events (ie. the inevitable chemical path that nature would otherwise follow). We demonstrate this when we use natural materials to construct, transform or destroy things from buildings to aircraft; bombs to art; music to scientific instruments, etc. Many will see our influence as a source of spontaneity, randomness, or design – ie. the very things that will plug gaps in theories of origin if only we were there 13.77 billion years ago.

At this point most people will assume that **if** the Big Bang was a moment of origin and it has taken a long time for living beings to evolve, then the only being who could possibly be present at that time was God. However that is not entirely correct because while God is a possibility, (I have never seen anything to disprove God), this line of thinking focuses on just one perspective and ignores the rest.

Most theories of origin assume that there was an eternal existence, and that the Big Bang was a natural moment of transformation within eternity - not pure creation. Physical beings could easily have existed in a prior Universe.

Being alive in the run up to a Big Bang

Because we exist, materialists and atheists have to assume that life, (as a mechanical process), would emerge naturally **in *any* environment where the conditions are right**, even though we don't know what those mechanisms or circumstances might be. If not, the emergence of life would be **entirely dependent on luck or an outside factor**.

For physical beings like us to exert influence over a formative event like the Big Bang, they would not only have to exist beforehand but must also have enough time to analyse, design and implement ways to manipulate such events.

They wouldn't have to resemble us in any other way, however the configuration of existence would have to be in a form that supported life. As science believes that there could be many other forms of existence which *couldn't* support life, then their emergence couldn't be guaranteed without luck unless there was a mechanism to put a suitable configuration in place. That mechanism is normally envisaged as a cyclical process that could try-out every permutation of existence.

Put another way, the main circumstances that would **prevent** sophisticated life forms from arising before the Big Bang event would be: a 'start point' for Matter/Energy (pure creation);or eternal existence without a cyclical mechanism, however there are only a few theories of origin which are based on these circumstances.

If there was an eternal cyclical existence then the fact that we are here is a strong indication that the right circumstances for life are **likely** to have also emerged in one or more of the infinite number of prior Universes that an eternal cycle would have generated. Within the lifetime of a suitable Universe there would be plenty of time for those earlier forms of life to develop sophisticated and powerful capabilities.

> Given the pace of human development over the past 200 years, such civilizations could be very powerful indeed within a very short time, (compared to the age of our Universe).

Of course, there may not have been an eternal cycle, but if we suppose that such a civilization did emerge in a Universe that existed before the last Big Bang, and they found themselves within a cyclical process that threatened to end their existence, then how could they position themselves to influence such a massive event?

Ways for Physical Beings to influence a Big Bang Event

As there is no way of knowing what might influence a Big Bang, this section will focus on 3 other pivotal issues which assume that such influence is possible:-

- The timing of an intervention
- An approach to intervening, and
- Clarifying what may need to be achieved

On the assumption that an event on the scale of a Big Bang is likely to overwhelm and transform everything within the visible Universe, (in whatever way that is achieved – say a super hot blast), then physical beings will probably not survive the event.

This would be especially true if we follow traditional expectations that a Big Bang would also generate a new configuration of existence – a new version of reality with different Laws of Physics.

> Although it can be argued that an advanced civilization may find a way to protect itself from the effects of such a transformation, (which may allow them to influence things during the event), this is generally regarded as being highly unlikely given the power of a Big Bang and may also be strategically pointless as it would probably isolate them forever if the rest of existence was now in a different and incompatible configuration.

It is therefore highly likely that the influence of such a civilization would only be exerted before the event – especially when we consider that the technology which they had beforehand is unlikely to operate under different Laws of Physics.

Pre-emptive action assumes that these beings could see the event looming, (whether that is in the final stages of a Big Crunch where a singularity was being formed, or in seeing that curtains of energy were approaching each other; or that huge fields of influence were interacting, etc.). This would give them time to consider their options.

When you compare the size of any civilization to the size of a galaxy, let alone billions of galaxies, the scale of any intervention which they might achieve seems less than a pin prick, but we don't know what they might be capable of, or what indeed might be required to achieve significant effects.

By way of example, if every person in existence was given a hydrogen bomb, human influence may be enough to completely destroy the Earth but unlikely to destroy our sun. Yet if beings like us were able to exploit natural mechanisms in the Universe and even start a chain reaction of sufficient power, who knows where it might stop?

The things which living beings 'bring to the party' are a capacity for design as well as a means of introducing fundamental change. However, as they are not Gods, they need a mechanism to deliver the changes, like a chemical process, a Framework for Existence, or punching holes in the fabric of space. In effect they would be getting the Universe to change itself based on their instructions. While this presumes a lot, we will consider the viability of such scenarios in the paragraphs below.

Triggers can be miniscule in comparison to the effects that they generate, and again, we have to be clear about what we are expecting this civilization to achieve. They may simply wish to survive the event, but regardless of whether they were successful or not, **our** expectations are likely to be rather different, even if we wish them well.

We simply need them to leave the Universe in the configuration that we find it today... which of course is a reality not a hope. However the nature of any intervention by them may give us a pointer on timing, capability, and any reliance on luck.

In broad terms, if their version of existence was different to ours, any planned intervention by them would effectively lead to a designer Universe – ours, yet that would be a strange choice for them to make given the consequences we discussed before, (their destruction or isolation). The precise motivations are irrelevant as long as there are some credible reasons for them to act that way... for instance, they might choose to act through altruism (ie. knowing that they couldn't escape their own deaths but could make things better for living beings next time around).

In terms of what they could do, we might speculate about a 'poison pill' that was fired into the midst of a growing singularity to alter its 'chemical composition' and therefore what it might achieve. However the nature of a singularity is that it will break down all matter into a common underlying substance, so it's difficult to see what change could be brought about. Put another way, the strategic difficulty with any chemical/nuclear process is that it should always work **within** the Laws of Physics and shouldn't change them, so it is difficult to see how any chemical/nuclear reaction could alter the fundamental configuration of existence.

Yet there is a more subtle issue as well. Regardless of all the research that they may have done, it seems impossible for them to have had any prior experience of a Big Bang event and therefore it seems highly unlikely that their desired outcome could be guaranteed. We would still be reliant on luck.

Equally, if they triggered such an event by accident, we would again be reliant on luck.

So, as a bit of fun, and to introduce ideas about a possible mechanism, let's consider how events might play out on this basis. Let's suppose that there **was** such a thing as a Framework for Existence and that it was possible to insert new parameters into it.

> The simple scenario is that a mad scientist had identified an ability within physical matter that was in fact an Internal Framework for Existence, but he/she hadn't fully realised what it was, or how far its reach extended.

> Let's say that this person wanted to test an idea about how to change stone into gold by deploying some type of messenger particle that he/she had engineered - effectively the code for new parameters of existence.

160

We could speculate that the particle would not only affect the material being tested but also tap into the *instant communication* mechanism of the Universe, inadvertently triggering the Big Bang because the scientist didn't realise how quickly those new parameters could spread.

In this scenario an individual on our scale might trigger the Big Bang and be totally overwhelmed by it, leaving no trace of him/herself or the environment in which their civilization previously existed, yet they would have helped to configure the next reality, (all-be-it inadvertently), so their influence would have carried-over into our universe. However that society couldn't guarantee our outcome if the effect was generated by accident, or even by design.

In overview we would be saying that thinking beings could be the source of new configuration settings for existence if the Universe already had a mechanism to implement them. Greater planning might be applied if the scientist had been wiser and more alert to the possibilities... but this again makes us dependent on luck concerning who made the discovery.

However there may be a better and more credible scenario – that the version of reality which the earlier civilization evolved-from is exactly the same as ours.

If the Universe was following an eternal cycle of change and this civilization could anticipate their own destruction, they might intervene to preserve the 'status quo'; to preserve themselves and their environment by breaking the cycle – stopping the next Big Bang. They wouldn't need to gamble on a new design for existence, and may be able to do something far less radical and more predictable.

As an example, they may find a source of power to accelerate the expansion of the Universe beyond the point that it could collapse again - preventing the next Big Crunch from happening. We might speculate about them discovering enormous reserves of energy in other dimensions and finding a way to release it in a controlled way across the entirety of the Universe.

Of course if that had happened it would have occurred after the last Big Bang... but we would need to allow time for heavy chemicals to emerge and then life to form... so approx. 9 billion years after the main event. Actually... that rings a bell somehow... ☺

I am not saying that this is how events actually unfolded. I am merely trying to illustrate that there **is** a potential for physical beings to intervene by exploiting natural mechanisms. In this case their contributions would lie in the creativity in analysing the deeper recesses of existence; developing the expertise to break through the different dimensional barriers (whatever they may be); and in their decision to act, (all of which is based in Thought, spontaneity, etc.).

> However in the absence of relevant natural mechanisms which existed on the scale of the Universe any beings below the ability of Gods would be powerless/impotent in this respect.

So in overview, ending the cycle in order to preserve what had already arisen naturally, overcomes the need for luck; whereas attempting to change the **next** Big Bang could leave us dependent on luck.

Much depends on whether you feel that creativity and Thought will produce an inevitable outcome because they could be based entirely in Matter/Energy.

Is Physical or Pluralistic Thought compatible with an Eternal Big Bang Cycle?

There are some subtle points in this section and in order to make them, I have to repeat myself a bit. I'm sorry about that, but I ask you to stay with it.

Philosophically there's a potential dilemma if we suggest that fundamental change can be brought about by physical beings, (including the ending of an eternal cycle), because such actions require spontaneity, randomness, or an **ex**ternal influence, yet ***Physical Thought*** cannot provide any of these things because it would be based on Matter/Energy within the pot of existence. How could they act in this capacity?

> The answer may depend on your point of view.

> Some people will say that they can't – which must either demonstrate that they have another element underpinning their existence – the underlying stuff of Thought, (making them Pluralistic or Intangible beings instead), or there is an **ex**ternal influence such as God.

> However there **is** an argument which Materialists can deploy, which suggests that Physical Thought is somehow out of sync with the rest of physical Matter/Energy and is therefore able to act as an **ex**ternal influence.

It has also been suggested that there is a second systemic problem with any notion involving the intervention of physical beings.

To avoid luck in generating our configuration of existence within an inevitable process of cause & effect, people have imagined an eternal cycle of change/experimentation. However if living beings are part of that process then anything they are capable of must also be a possibility, and if that includes ending the cycle, then the cycle itself could **not** be considered eternal. At headline level this 'mechanism' could not fulfil its prime purpose, so no scenario involving physical beings could therefore be viable.

However that is not the full story in two basic respects.

Firstly, thinking beings are intelligent and can design outcomes related to their abilities in manipulating the physical environment. Put another way, design can potentially avoid aeons of experimentation by guiding nature directly to the desired outcome. It doesn't need eternity to do this as long as nature provides the necessary mechanisms/facilities.

Secondly, the prime requirement is to achieve our existence exactly as it stands today. There is no requirement about what happens next, so there is no requirement for an eternal cycle to continue afterwards.

Put another way, the relevant cycle needs to be ***potentially infinite*** to allow enough time for our existence to be achieved but it is equally likely that the desired result may be achieved towards the beginning of that range, as later in the process. So the additional refinement we need to the basic scenario is that **any end to the cycle must be guaranteed to *only* occur after our configuration is achieved – not beforehand.**

On this basis I'd like to pose a question.

Would the emergence of a sophisticated civilization **inevitably** mark the end of an eternal cycle because they would always want to take control, (possibly as a way to survive and thrive), by preserving or tweaking the configuration of existence?

It can be argued that if a civilization didn't manage to take control, the eternal process would continue to the next Universe, and so on until such time as a civilization did take control. The counter to this is that living beings may make a mistake - ending the cycle before it reaches the desired effect, (perhaps because they don't know all the factors at play, or just through incompetence).

However, let's suppose that the **only** way to end the cycle would be for a competent intelligence to know **all** relevant circumstances, *then the emergence of sophisticated life may be a way to guarantee that the cycle is only ended once the desired outcome is achieved.* If so, we have an interesting scenario.

Existence would almost seem to be set up to achieve life with a certain level of sophistication, which would then preserve itself in that form.

I find this to be a very intriguing idea, but in fairness to other perspectives living beings may not be the only way that an eternal cycle might be brought to an end.

However as a way to retain the involvement of physical beings, their fallibility (in terms of lack of knowledge, or even their propensity to make mistakes), can be tacked from another angle, by challenging the need for zero risk, (the total absence of luck).

Would a 'good bet' be adequate for the purposes of providing an explanation? Would favourable odds of say 80:20 be good enough? That depends on your attitude to the principle of causality. At its core, '*single cause : single effect*' can only provide an inevitable solution which would provide 100% certainty... however in terms of our explanation, can we assume that exactly the right circumstances can be implied if we get 80% right – saving us having to guess the residual 20%?

A personal choice.

On the other hand, if a 2nd type of stuff has always existed, (sitting alongside Matter/Energy), and it is capable of generating spontaneous or random effects, how could any physical process be expected to be eternal? The other stuff would be constantly 'putting a spanner in the works' so no physical mechanism could explore every permutation of existence.

This is another real dilemma for certain theories, because we may need spontaneity/randomness to generate fundamental change, yet it could easily lead to instability or even retrograde steps.

Clearly we don't know enough to answer that question when the existence of such stuff cannot be detected – only inferred. In terms of speculation however, the whole basis of Dualist philosophy is that change can be introduced in a way that doesn't result in chaos, and one way to do this may be to always filter a spontaneous idea through an assessment based in Thought.

However the counter-argument suggests that there is no obvious reason why such stuff shouldn't influence physical Matter/Energy directly, by-passing living beings altogether, which would make concepts of eternal processes less likely.

Dualist theories of origin have to explain how a 2nd type of stuff would not lead to chaos at all stages in the development of our Universe.

11.3   <u>Pure Creation without God</u>

Our consideration of pure creation will involve some mental gymnastics, so hang onto your hats! ☺

As we have seen, the basic logic of prime origin is that if existence has not been eternal, (without beginning), then it must have arisen from pure creation, which is the only way to provide a beginning in the absence of anything else.

Pure creation increases the stock of existence, and the basic scenario is that where there was nothing, there is now something.

This would be extraordinary in itself, but to mathematicians and physicists this scenario poses a deep problem because it fundamentally challenges a core principle – that the equation of existence must always balance. For this reason people who believe in pure creation as well as mathematics must try to find a way that can allow pure creation to balance the equation.

Typically they manage to do so using a mathematical sleight of hand.

If it can be demonstrated that new creation always produces balanced opposites (whether matter & antimatter, or positive and negative stuff), then mathematically the sum will add to zero and can therefore be said to balance, (ie. what was zero before, still adds to zero now).... and yet the pot of existence will have increased.

The best way to get our minds around this is to imagine that a black and totally empty void suddenly contains two pools of gleaming stuff, that are equal in size, but one is 'positive' and the other 'negative'. I think that most people would agree that the stock of existence had increased because there are now two lumps of something when there was previously nothing, however mathematically the two pools still add up to zero.

To this extent, when trying to assess whether there is existence or not, I would say that mathematically we should ignore any positive or negative status as the stuff would be there regardless. However if you're determined to balance the equation, balanced opposites can be applied in different ways.

In Chapter 9 we considered the suggestion that "nothingness" could be **split** into matter and antimatter, however this **didn't** represent pure creation. The suggestion that something can be split means that it has to exist in the first place – so this would represent a transformation of something eternal which we couldn't previously detect.

Pure Creation can only be said to occur if it comes from absolutely nothing.

The pure creation version of the zero equation is therefore to say that stuff suddenly appears as balanced opposites, but it wasn't the result of a split in anything.

However this doesn't make it true. We deem our environment to be made of 'positive stuff' but we don't know of any pools of negative stuff, (ie. 'antimatter'), to balance it. The evidence available therefore appears to support the traditional notion of pure creation which only produces 'positive matter'.

This lack of evidence for large quantities of negative matter is a real problem for those wishing to preserve a balanced equation, and their assumption is often that 'it must be there' it's just that we haven't realised that some of the things we see represent the negative stuff. As an example of some lateral thinking, if the underlying stuff of Thought provides spontaneity & randomness which are the opposites of cause &

effect, then could the stuff of Thought also be the opposite of Matter/Energy – the negative stuff?

Others have suggested that if positive matter exists in the outer parts of a galaxy which we occupy, negative matter may be grouped in the middle to balance it. In other words, a black hole may represent the elusive pot of negative matter. However this is only one suggestion.

Whether or not there are balanced opposites, pure creation must still come from nothing because it must increase the pot of existence. This doesn't change even if it arises next to, (not too far away from), something that already exists.

Without any idea how pure creation could work, but in the knowledge that it doesn't seem to be occurring everywhere all of the time, the next assumption is that it would be dependent on the right circumstances arising even if spontaneity is required as well.

Unlike other forms of Big Bang mechanism where reactions arise from enormous things like a singularity, pure creation is generally thought to arise in a void rather than *within* something that exists. As we saw in Chapter 10, that may not be an easy thing to achieve because a number of things seem to occupy all parts of space: from the cosmic background radiation, to particles of light, dark energy, and dark matter.

On this basis we might speculate that 'generating of a patch of absolute void' might be one way in which the circumstances for pure creation could arise.

The purity of the void may increase it's potency, (ie. its potential to generate pure creation), not by any sort of pressure difference with the wider environment, but by making it more likely that this special type of spontaneous event might arise, (ie. pure creation may not be able to occur where something already exists).

In the absence of any other logic for the 'mechanism' of pure creation, and what it might generate, (Matter/Energy or another type of intangible stuff), other speculation about 'creating the right circumstances' tends to be more abstract.

- Would the 'right circumstances' purely relate to **physical** conditions or something else?

- Would pure creation be an **inevitable and automatic** reaction to the right circumstances, (ie. it happencd immediately), or would it additionally require some other sort of spontaneous action ?

The reason for posing these questions is that if a void *inevitably* generated pure creation, no void could last more than a moment, so any suggestion that a void may have been a natural, prolonged, and possibly eternal state would either require an *external* trigger or a spontaneous act (without cause), to change it.

As it is very difficult to produce a perfect void we may have one way to explain why there is no evidence for other acts of pure creation, but this presumes many things - not least whether the observation is correct.

It's quite conceivable that Pure Creation may be happening all the time but on a small scale that is either below our ability to see it, or which we see but attribute to other things, such as exploding supernovae.

Clearly if this is correct then we would have to wonder why the Big Bang was such a dramatically different scale of event.

In truth, without having any clues to what is truly possible this is the limit of our speculation about triggers, so let's consider a more basic issue: what is pure creation likely to produce, and how can this lead to our Universe without resorting to luck?

We should remember that the universe contains a vast amount of physical matter, (close to 200 **billion** galaxies), that are relatively evenly spread, but whose speed increases with the distance that they are away from us. They form an even pattern.

A myriad of separate pure creation events would struggle to create this pattern – especially as these events would seem to be distinctly individual and un-co-ordinated.

But co-ordination is not the only issue, We have to consider what would emerge directly from a creation event.

Without the influence of God I suspect most people **wouldn't** believe that fully formed cars, people, planets, or galaxies were likely to emerge directly. They would probably expect the raw stuff of existence to be produced in line with scientific thinking, and through a process of change this could slowly develop more sophisticated features.

The direct emergence of anything sophisticated, (and especially a fully functioning world), would smack of design, and without God that would be hard to explain. But to our knowledge, scientifically, it doesn't happen.

If pure creation generates material in a very raw state, it may do so as a drip feed, or a Big Bang. Most people will naturally think of it producing Matter/Energy but of course, if there was more than one type of stuff it may produce either. We then have to face the fact that both types may have the potential to adopt many forms.

When we ask the question 'which configuration of existence the new material would adopt'? many will assume that because we don't see different forms of reality scattered across the Universe, any new creation **within** our Universe would adopt the same form as everything else – but unless there was a mechanism like a Framework to enforce standardisation, there is no apparent reason why it should do so.

An alternative may be that different configurations would compete in some form of self-balancing mechanism and thereby be transformed into a common standard.

If we consider the scenario that absolutely nothing existed before the Big Bang then a Framework couldn't exist before the pure creation event and so it would have to evolve very quickly in order for it to standardise the form of everything else. To me this seems less likely than the emergence of a self-balancing process of some sort... but who knows if either are possible?

There is also the possibility that pure creation on the scale of a Big Bang could still occur alongside something that already existed. In this context logic suggests two other main ways to explain the uniformity of existence:-

- that the last creation event overwhelmed everything else, either by physical force or through a chemical process that converted everything to the new standard, or

- the newly created material pushed all pre-existing stuff beyond our gaze, or

What we **can** probably say is that no single pure creation event which spreads its outputs everywhere could be infinite in size/scale, in the same way that no explosion could be infinite. The only way in which I can envisage it being infinite is for it to occur everywhere at once.

In this respect it's interesting to remind ourselves of the findings from NASA which lead it to believe that space isn't curved[1], which means that outer space is potentially infinite in size. Yet the significance of this partly depends on whether you believe that space has to be created before it can be occupied.

From all of the above, we can see that the notion of pure creation is one which presents opportunities and challenges – not least, whether it is possible at all.

The very idea of something coming from absolutely nothing seems impossible, but that may be a result of our conditioning. Just because we can't see it doesn't mean it isn't happening, and in the case of new thoughts popping into our heads, there may be more evidence than we initially suppose.

Stephen Hawking initially entered the debate about origin with his fabulous book, 'A Brief History of Time', where he left the possibilities for 'God or no God' open either way. Yet his more recent contributions have sided firmly with the suggestion that God is not necessary and everything comes from the intrinsic properties of matter/energy... and possibly antimatter.

A more recent work with Leonard Mlodinow, is entitled "The Grand Design" and in it he makes some controversial points, as evidenced in the following quote:

> "Because there is a law such as gravity, the universe can and will create itself from nothing... Spontaneous creation is the reason there is something rather than nothing, why the universe exists, why we exist. It is not necessary to invoke God to light the blue touch paper and set the universe going."[20]

By talking of spontaneous creation a lot of commentators interpreted Hawking & Mlodinow's comments to mean that the singularity at the core of the Big Bang suddenly appeared from nowhere out of absolutely nothing from an act of spontaneous pure creation, without time.

If he's being interpreted correctly, that's a big claim which also challenges long standing basic principles of science based on cause and effect. I also find that it's a surprising claim as it dismisses the possibility that physical matter has existed for ever, (without apparent explanation).

Yet my own reading of their book suggests something slightly different – ie. the spontaneous creation of balanced opposites, but instead of talking about matter, they refer to 'energy' and speculate that the 'total energy of the universe must always remain zero'. In this context they suggest that if positive energy is required to form physical matter, then an equivalent amount of negative energy must exist as well, and they propose that this manifests itself in a 'negative' effect - gravity.

Of course, we don't know what gravity represents, but it's a nice idea to think that if gravity is proportional to mass, then the two must somehow balance out. The idea also has some parallels with the notion that antimatter has largely moved to the centre of each galaxy where a supermassive black hole may have gathered most of the negative energy in order to hold all of the solar systems in place around it despite the enormous distances involved.

It's a nice twist on the earlier permutations.

However at face value the notion still requires something to emerge from absolutely nothing. As we saw earlier, the fact that it balances to zero doesn't change that, it only means that the purity of a balanced equation is preserved.

The only alternative is to go back to the old notion of 'splitting the zero form of something' – ie. splitting some form of eternal stuff that exists but which we can't detect in its zero form.

In addition, by implication, the Big Bang would be a one-off event which probably means that we are entirely reliant on luck to achieve our outcome unless there is only one way for existence to emerge – ours.

Yet for many people, we come back to the fact that Pure Creation is something that our logic cannot rationalise – perhaps because we are so used to an environment which doesn't produce things out of nothing and does normally balance the equation.

It doesn't matter whether it's triggered by God or some spontaneous act of nature, we simply cannot rationalise how something could come from absolutely nothing.

This is even more true if we confine existence to physical Matter/Energy which doesn't seem capable of spontaneous acts, and we deny the types of Thought which may be able to do this.

So the next section does consider how the spontaneous creation of Thought might translate into physical existence.

11.4   Squaring Religion with Science - the Influence of Thought & God

> The only proof needed for the existence of God is music.
>
> (Paraphrasing Kurt Vonnegut)

Concepts about God can take many forms, yet in the context of this debate we are simply regarding God as a way to produce our Universe.

> As readers generally find it very tedious to have constant references to God as He/She/Them, (even though this would be fair to the different perspectives), I will use the convention of referring to a single Creator God as a 'He' because it keeps things simple and is the dominant view in most communities, but this is not intended to deny the other possibilities.

In terms of explaining prime origin, the central role of a Creator God is to create, and when discussing this I believe that most people will interpret it to mean that He had performed the miracle of producing something from absolutely nothing... pure creation. However this is not the only option which a deity might pursue, and various scriptures refer to God shaping an eternal existence – bringing order out of chaos.

So the first thing to realise is that the underlying strategic options for our origin which include God are the same as those without Him; ie. pure creation, transformation, or a mixture of the two.

> If you think that resorting to 'transformation' somehow demeans God, (and that 'pure creation out of nothing' is the only acceptable format for Him to adopt), then I suggest you try forming something solid out of the energy of a fire using nothing but your mind. Then try to imagine your lack of success on the scale of a universe.

Now try telling me that transformation is not impressive even by God's standards!

168

An advantage which God brings is that He can effectively guarantee that the right outcome will emerge; in other words He can trigger things when no other trigger could exist, and He can overcome luck by shaping the outcome.

Most religions will also position God within an environment and the range of beliefs cover various possibilities, such as :

- a complete void in which He is the only thing to exist;
- a realm made of chaotic and therefore unstructured background 'stuff';
- more elaborate notions including seas, chasms, air, cosmos, etc.

In all cases God has the ability to manipulate His environment, but it is generally unclear whether He is made of the same stuff as His environment, or represents something fundamentally different to it.

Some people argue that God **is** existence – ie. He is the very stuff of everything that exists. Others say that He is a being **within** the totality of our existence, while there are those who argue that God is separate from and therefore **outside** the totality of our existence. Finally, there are those who say that God doesn't exist at all, however they have had a fair hearing already. The purpose of this section is to consider what may be possible **with** the presence of a God.

For the most part, God is portrayed as a being who is extremely powerful, extremely aware, and either having extreme longevity or is immortal. He is therefore perceived to be allied to Thought with all of its potential properties of design, creativity, spontaneity and randomness. Some definitions go further to say that He is either present everywhere, or His influence can be felt anywhere instantly.

Yet within the range of possibilities there are also minimalist concepts which simply define a creator as 'whatever it took to get dynamic existence started' and some of these concepts don't portray God as a being.

When considering the strategic ways in which God might intervene in an **existing** Universe, He might add-to, take away from, transform, or enhance/degrade something that already exists. The general logic is that if He had created something He would want to use it and work with it, rather than start afresh each time He wanted to do something. On this basis there is a rationale about why He may wish to enhance nature's own mechanisms, from induction, to vacuums, self-balancing mechanisms, and rips/channels in space.

God may be able to do anything, but we have to separate notions of Him creating something which has new properties, from scenarios where He works with the things He has already created and therefore within their normal abilities and circumstances.

We also find that **definitions** can place limits on what He might do – not in terms of restricting His infinite abilities but recognising that if He wants something to be categorized in a certain way, then He would have to comply with that definition.

For this reason His ability to intervene becomes slightly more complex when we consider Pure Creation.

It's worth reminding ourselves that for an event to be labelled as 'Pure Creation', the things that were created must simply appear - they cannot be made of anything which existed beforehand, including God's stock of power.

As mentioned earlier, new things might simply pop up alongside the old, and may even combine with them almost immediately, but the created elements themselves cannot be made of anything which existed beforehand if they are to increase the stock of existence. This has implications for any God-based explanations.

So we can begin to sub-divide the options for creation by God into:-

a) transforming part of Himself into Matter, (if He was originally the only thing to exist), or

b) transforming things that are distinct from Himself into Matter, (either His environment, or a stock of power/energy which He has built-up), or

c) causing Matter/Energy or Thought to simply appear out of nowhere – leaving His own power untouched in this respect : pure creation out of nothing

All of these hold true in both the Dualist and Idealist perspectives.

To believe in a creator God you generally have to accept that something (Him) did exist before the moment of creation, (eg. the Big Bang), which is the essence of the Cosmological Argument.

If God spontaneously emerged at exactly the same time as physical matter/energy then He couldn't have created it, (ie. He could not be the cause), but He could still transform it later. So on the logic above there are limits to how God can operate if He is to meet the definition of pure creation.

***He may create the right environment for pure creation to occur,
including His command, but little else.***

The only caveat to this that I can think of is that while creating the right circumstances for pure creation God may be able to incorporate a design, which would mean that the outputs of pure creation emerged in the configuration that He wanted, rather than Him having to transform the new stuff immediately after it came into existence. While this may please academics I personally see little practical difference between the two.

Many people will equate the Big Bang moment with the point at which God engineered our Universe however as we weren't there to witness the event, there is no way of knowing what God actually created.

The range of possibilities run from a fully formed complex end product, to a minimalist approach which just provided the raw stuff of existence with a set of basic configuration settings and a way to evolve. We saw a list of possibilities in Ch.8 which include the 6 main configuration settings for existence, and the possibility that He was the creator of a Framework for Existence to give Him instant control anywhere in existence.

It's a viable explanation which means that it falls within the range of options, but again that doesn't necessarily make it true.

Personally I think it would be rather clever and subtle of God if He designed the mechanism of evolution and could predict how things would change to produce our outcome rather than Him having to work through an enormous list of individual complex creations. If each 'God-Day' represented 2 billion years in our time, He could even do it all in 7 days! However, as always, there is a range of opinion.

Yet if we assume that pure creation is possible, we should ask if there is any other God-related concept of how anything might emerge from absolutely nothing? Well if you remember the range of philosophies about origin, you will probably be able to guess some of the methods I'm about to present.

If you believe that everything is in the mind, (as in some forms of Idealism), then perhaps new thoughts **can** simply emerge, in the absence of true substance. As already mentioned, we do see our pool of ideas growing daily, and yet we **don't** believe that we're converting any 'stock of blank ideas' into 'real ideas'.

The major benefit of such an Idealist philosophy is that pure *physical* creation would **not** be necessary, as the physical world would just exist in people's minds. In this context God could represent an overarching framework for our thoughts, and therefore any perception of our reality might change instantly at His will.

In other words, He could simply dream-up a new idea and circulate it to us in order for it to become our reality. The powers of such a God would be truly unrestrained by physical capabilities as there wouldn't be any need to manipulate a separate physical environment. At worst He would 'just' have to re-write the rules which regulated our thoughts, like re-writing a computer program.

The Idealist concept places God **within** the totality of our existence, but it isn't a popular idea because most people will consider that physical matter is real and does have a true independent and substantial existence. (In part this may be because our sense of reality is a lot stronger than our dreams, which is why we separate the two realms and give credence to physical substance).

However, if there was originally no physical existence, just His own thoughts, then **He** would presumably be made of a different type of stuff – perhaps the stuff of Thought, so if true substance does have to be created from these beginnings then we have to accept that a deity would have to do something remarkable.

He either had to transform the intangible stuff of Thought into true physical substance, or He had to get pure creation to produce something that had never existed before.

You may suppose that pure creation would be more likely to produce stuff in the same form that existed at the time (ie. more Thought), and we should remember that the strongest example of pure creation which we currently have lies in the growth of our new thoughts. So the generation of physical substance is a true challenge.

To gain some concept of how this might be achieved we have to consider any touch-points there might be between Thought and physical reality.

11.4.1   The Emergence of Energy/Power from the Stuff of Thought

This brief section is highly speculative, (probably even more than the subject of God), but it is included here for the sake of completeness and to give you some ideas which you may find useful in other contexts.

If there is a second type of stuff which underpins our existence, it is entirely possible that this came into existence first and not Matter/Energy. This stuff would either be eternal, or was the result of a first bout of pure creation.

If the only thing to exist was the stuff of Thought, then in order to generate physical substance we would either have to say that pure creation could produce Matter/Energy next time it occurred, or that the stuff of Thought could be transformed into true physical substance – the intangible somehow becoming tangible.

While some might say that this conversion simply happened at the command of God, others will look for a solution within the properties of the world that we occupy, even if it is only to provide an outline approach. This is where it can be useful to look at potential touch points between the two types of stuff.

Perhaps the most obvious link is that Thought has the ability to influence Matter/Energy, because our minds control our bodies. On this basis there must be some common ground that allows our thoughts to be translated into physical actions.

Science has determined that all activity within the physical environment has to be based in energy, which would therefore imply that some form of energy transfer must occur between the 'intangible' and physical realms. This would be our touch point.

The moist obvious suggestion is that the stuff of Thought also contains a certain amount of energy. No matter how small that may be, if you built up the number of thoughts you would be building up an underlying level of energy. It may be a different type of energy allied to a different type of stuff, or it may be the same energy used by the physical world, but for effects to be generated in physical matter there needs to be an exchange of energy.

One way in which this exchange might occur is through an interaction between 'fields of Thought' and 'fields generated by physical matter', possibly resulting in an induction effect which could allow energy to transfer between the two realms. However this wouldn't explain how matter was first created, because before physical Matter/Energy existed only one type of field could exist.

We therefore need another approach.

One theory is that after a long time, pure creation of Thought, (the 2nd type of stuff), might have generated a sizeable stockpile of energy which reached a level where it discharged into space as the Big Bang, and in the process, was converted into matter.

So what has the concept of God added to this way of thinking? The obvious implication is that the spontaneous build up of new Thoughts might represent the mind of God and the associated build-up of energy forms the basis of His power.

Once again, it is up to you to decide how far to take those arguments.

12 A Route Map to our Existence

Well here we are at Chapter 12 where we set-out all of the options for our origin side by side, and **you** get to decide how the world and our entire universe were created. Wow, I can't wait to find out ! ☺

This is our 'cook-book', where the recipes, (Theories of Origin), must reflect what can be done with the available ingredients using the mechanisms that might fashion them into our reality. You'll be pleased to know that I won't be asking you to 'bake the cake'. That could be a tad difficult. Instead, our aim will be to judge how likely each process of origin would be to produce our **exact** Universe without resorting to luck.

By 'our exact Universe', I mean the bits that astronomers can see, rather than anything speculative/unknown. Our visible Universe may be part of something much larger, but this is the only bit which has to be explained. An *infinite* universe is beyond our remit.

All processes are defined by a starting condition, a mechanism, and possibly an enabler; and in these circumstances the good news is that there are only a modest number of combinations, so it is possible to work through all of them within an hour or two. This chapter will present a way of approaching such an exercise.

I will also provide you with one set of possible results to illustrate how the exercise might work and to give you a yardstick by which to make your own judgements. However it is important to realise that there is no fixed answer to this, and that each person is likely to reach different conclusions based on their own preferences.

The results I present show a small number of viable possibilities and it's interesting to see which ones emerged strongest. The full rationale for each of my assessments can be found in Appendix A for those interested in delving deeper.

As a reminder, let me present the diagram which summarises the task at hand.

The Conceptual Big Bang Process

State of Prior Existence	Mechanisms	Intermediate Needs/Enablers	Final Outputs (our Universe)
No Physical Matter - Total absence of anything - Only energy/chaos - Splittable nothingness	Spontaneous Pure Creation Release of Energy - God (poss. by direct creation) - Physical Structures Splitting of Nothingness	Triggers Configuration Settings Processes & Sequences	Avoidance of Luck Even spread of galaxies All galaxies appearing to move outwards
Scattered Matter - Static - Dynamic	Induction through Fields Special types of Vacuum	Framework Time	Uniform Configuration Size within Timeframe
Concentrated Matter - Static - Dynamic	Self-Balancing mechanisms Chemical/Nuclear - single blast - prolonged chain reaction Direct Transfer - only - with transformation New Parameters Inserted into a Framework for Existence		Dynamic & Reacting Able to Control & Regulate itself. Capable of Life

Although this analysis will use the term 'Big Bang' as a metaphor for the start of our Universe, we should remember that this event could represent a variety of processes.

12.1 <u>Matching Mechanisms to Key Factors</u>

In broad terms one element from each of the first 3 columns in the diagram above must be combined to form a single conceptual process which we can then assess for viability.

In fact, having gone through the exercise, I can tell you that the most significant factors in making these assessments will be how *mechanisms* match the possible *starting conditions* to produce our universe. I found that the impact of 'Enablers' was generally absorbed within the starting conditions as they are closely linked.

Yet it would be wrong to assume that all of the items listed are compatible with each other. Both the tables below, and those in Appendix A, demonstrate that the vast majority of combinations are either not viable (coded black) or highly unlikely (coded grey), and this serves to considerably narrow our range of real possibilities.

However we initially have to give each one a fair hearing. The 70 basic scenarios are shown in the table below, reflecting combinations of 6 starting conditions (cols. a-f) and 12 mechanisms (rows 1-12); one for each cell in the grid, less two duplicates.

Although I realise that some people can feel daunted by the prospect of a table please be reassured **we are just looking at colours at this stage**, and on this basis a data table is the clearest way of showing the different mix of factors which we have to cover. I have done my best to simplify the presentation, and it **will** be informative at the end.

Starting Circumstances – Matter/Energy Alone

		a	b	c	d	e	f
	Black or Grey = not viable	**Total Absence**	**Absence of all Matter**	**Scattered Matter**		**Concentrated Matter**	
	Red = Unlikely - 2 major problems			Static (eg. without Time)	Dynamic	Static (eg. without Time)	Dynamic
	Yellow = Possible – 1 major problem	Nothing physical in our 3 dimensions	permitting unstructured energy & forces				
	Blue = strong possibility / Likely						
Ref	**Mechanism**						
1	Spontaneous Pure Creation						
2	Creation by God						
3	Energy Release - Structures						
4	Splitting Nothingness						
5	Induction						
6	Vacuums/Voids in space						
7	Self-Correcting Structures						
8	Chem./Nuclear - 1 explosion						
9	Chem./Nuclear Chain Reaction					N/A - Duplication	
10	Direct Transfer - Channel/Tube						
11	Direct Transfer - Dimensions						
12	Frameworks & Parameters						

The way in which I have assessed viability is to determine the number of significant problems that each process contains, which also means that the theories can be directly compared, even though they are very different. The logic behind this approach is that they each have their strengths & weaknesses, and while we might have endless arguments about which unresolved issue has greater significance I have taken the view that even one serious point which remains unanswered is a challenge to viability.

At the same time I recognise that humanity still has a lot of gaps in its knowledge, and some of these issues are likely to be resolved in future. While some issues may appear more insurmountable than others, we simply don't know what the future may reveal, yet we can say that the more issues there are, the less viable the scenario becomes.

This is what my grading (colour code) reflects.

No theory seems able to prove the existence of its key mechanism(s) so we ***can't*** focus on the ones which are 'real' because they **all** have issues of credibility, and if they all suffer in the same way, lack of evidence is not a way to distinguish between them. Eg.

> Vacuums and Chemical/Nuclear reactions have not been shown to generate pure creation, and always seem to work *within* the known Laws of Physics, not change them, so they **don't** seem able to produce new versions of reality;

> Singularities, Curtains of Energy, Frameworks for Existence, other dimensions, or the 'multiverse' have not been shown to exist; and effects such as Induction have never been demonstrated on a cosmic scale.

As a result I do **not** count 'lack of evidence for the key mechanism' in the scoring.

To this extent each theory/combination is on a level playing field, carrying the assumption that its mechanism is a possibility and may also be viable in the right circumstances. Yet, as mentioned before, not all combinations of starting conditions and mechanism are valid. As a simple example, you can't have a physical mechanism in the complete absence of anything.

If you disagree, with my 'level playing field' approach then of course you are free to focus on the mechanisms which you believe do exist and ignore the other rows, however the grading I have applied is as follows:-

✔	A strong or even 'Likely' possibility that has **no major issues** beyond the existence of the mechanism itself.
	A reasonable possibility which has **one 'challenging' issue** that may require an additional factor or luck to make it fully viable.

To help us focus on the strongest possibilities I have chosen to 'draw the line' at this point to show other colours as different levels of **un**likelihood.

	Unlikely to be successful because it has **two significant problems** with the process/scenario.
	Highly Unlikely or Not Possible due to **3 or more significant problems.**
	The combination of Mechanism & Circumstance is **not applicable** – ie. invalid / cannot occur.

However if you look closely at the first diagram in this chapter, you may see one particular factor that's missing from the lists. Thought.

With the exception of one mechanism, (Creation by God – Row 2), the diagram & table above reflect those elements which must be accounted-for *just* by Matter/Energy in its raw form. In short, the table above reflects an assessment based on the basic **materialist/atheistic perspective**, so Row 2 is inevitably shown in black.

Yet ALL viewpoints **do** acknowledge that some level of Thought does exist beyond the raw workings of Matter/Energy, (because we are all here, thinking), and therefore we can layer different degrees of Thought on top of the basic table before re-running the exercise to see what difference this new factor would make. I have considered 3 'levels' of Thought, reflecting the different possibilities outlined in Chapter 11:

a) the influence of thinking *physical beings* who emerged from Matter/Energy alone, and who therefore use '**Physical Thought**'.

b) the assistance of a second type of underlying stuff that brings spontaneity & randomness, as well as the potential for **Pluralistic & Intangible Thought**.

c) **Divine Thought**, guidance, and power.

So, overall, I have assessed the table above in 4 different ways, and this is what I summarise in the larger table below. Please take a quick look at it now so that you understand the explanations I am about to make. Those different types of assessment equate to different sections of the table made up of groups of columns (labelled a-f).

As all of these assessments presume that Matter/Energy is real, none of them reflect the 'Idealist' perspective. However Idealism is a relatively simple viewpoint, because any explanation of reality would boil down to a manipulation of Thought & ideas which would, in turn, mean that virtually anything is possible.

Before looking at the findings, we need to understand the exercise in a bit more detail, beginning with what the rows and columns in the tables represent.

The columns represent starting conditions, and column 'a' envisages a complete absence of anything – but possibly only within our 3 dimensions, (the parts of existence that we can access & occupy). If there are other dimensions of existence, (as proposed in M-Theory[22]), it may be possible for stuff to exist there, and be totally invisible to us.

Within column 'a' there are only 2 potential mechanisms (rows) which could fill empty space with our physical Universe, (excluding God). Row 1 is the most obvious, (pure creation), which might generate new existence from absolutely nothing. Row 11 represents the other approach; filling our empty 3D space by transferring material from another set of 'parallel' dimensions - the only other place where stuff might exist.

All other columns assume an *eternal* physical existence prior to the 'Big Bang' event 13.77 billion years ago, (whatever that may have been). Column 'b' represents a **perpetual** state of chaos - unstructured scattered material without any effective Laws of Physics. Columns 'd&f' presume that at the time of the Big Bang a dynamic structured environment did already exist with its own Laws of Physics, (although it may have been different to our own), while columns 'c&e' presume that this earlier physical existence was originally completely static, (without Time).

The only way to provide movement in an entirely static environment is through a spontaneous act, or an **ex**ternal influence. As a result, I have assumed that the presence of either a '2nd type of stuff' or God would enable static environments to become active, and therefore indistinguishable from their dynamic counterparts.

It is this logic which enables me to combine column 'c' with 'd', and column 'e' with 'f' within the summary table below, (to save space), once they are all dynamic.

You may also notice that in relation to Physical Thought, (physical beings), only column 'd' is presented. This is because a dynamic existence with scattered physical matter is the only circumstance where physical beings could arise.

In contrast, a '2nd type of stuff' that is able to generate Intangible Thought could be active in any column other than 'a'.

Columns 'e&f' deal with the notion of a singularity where all physical existence is gathered together in one place. Some scenarios (like Bang-Crunch) assume that singularities are also able to change the configuration of reality as a way for nature to explore every option, but there is often **no** explanation as to how this change might be achieved - which I regard as a weakness. We should also remember that an eternal cycle needs to preserve some Laws of Existence through every Big Bang event otherwise an eternal cycle of Big Bangs would not be possible, and to do this many scenarios need to invent other unknown factors of existence.

As already mentioned, within the rows dealing with chemical/nuclear reactions, '9e&f' (chain reaction in a singularity) effectively duplicate '8e&f' (exploding singularity), so I have excluded them to avoid distorting statistics that may emerge from the analysis.

I have also assumed that deities would not be able to exceed the powers/abilities of the stuff from which they are made, which may either be the same types of stuff that make up our environment, or something entirely different - beyond our environment.

While a Creator God may be able to do almost anything once He exists, His unlimited power is **only** reflected in Row 2 – Direct Creation by God.

In order to assess natural mechanisms for what they are, I make it a rule that if God changes any mechanism to force it do something else, it would represent a new creation that could only form part of Row 2. On other Rows He therefore has to work within the natural capabilities of those mechanisms and could only manipulate the *inputs* to the relevant process. In this context He therefore represents a way of bringing natural mechanisms to their full potential.

Finally in relation to this introduction, I have taken the approach that we are seeking prime causes. Any *thing* or event which follows-on from the initial action will **not** be the mechanism we are assessing – just a natural consequence of the prime factor.

So after the legalese, let's finally look at some results from the summary table below!

The first thing you will see is that the number of blues and yellows increase as you move from left to right. The increasing influence of Thought in its various forms is obviously kicking-in. Yet you may be surprised to see that the cells on the right, (reflecting the influence of God), are not all blue, and this is because God would be working **within** the natural capabilities of the mechanism or environmental circumstances, and not creating something new in those cells.

Another of the things you will notice from this table is the inclusion of 'L's representing the need for luck in some circumstances – even with God's intervention. This is generally because of the nature of the scenario, where *many* outcomes would be possible from the mechanism, yet there is no conceived way to guarantee our version of reality, and no ability to have multiple attempts.

'S's represent a need for spontaneity or randomness.

Summary of the Viable Scenarios for Our Origin

Legend: L/S = reliance on luck/spontaneity; ✓ (screw icon) = strong possibility / Likely

Ref	Mechanism	Materialist — Total Absence (a) Nothing in our 3D	Absence of Matter (b) chaos in energy & forces	Scattered Matter Static (c)	Scattered Matter Dynamic (d)	Concentrated Matter Static (e)	Concentrated Matter Dynamic (f)	With physical beings (d)	Dualist (2nd type of Underlying Stuff & Intangible Thought) a	Dualist b	Dualist c/d	Dualist e/f	With God a	With God b/c/d	With God e/f
1	Spontaneous Pure Creation	L/S				L/S	L/S		L	L	L	L	✓	✓	✓
2	Creation by God										✓			✓	✓
3	Energy Release - Structures		S								✓			✓	
4	Splitting Nothingness		S								✓			✓	✓
5	Induction				L			L		L	L			L	
6	Vacuums/Voids in space													L	
7	Self-Correcting Structures				L			L			L	L			✓
8	Chem./Nuclear - one explosion						L/S	L			L				L
9	Chem./Nuclear Chain Reaction				L	N/a- dupl'n		L			L	N/a		L	N/a
10	Direct Transfer - Channel/Tube														
11	Direct Transfer - Dimensions	L/S	L/S		✓		L/S		L	L	✓	✓		L	✓
12	Frameworks & Parameters						S				✓	✓		✓	✓

178

Overview of Viable Options

In looking back over the book we can see that philosophical logic gave us two basic options for the origin of existence – pure creation or eternal existence.

Looking at the top row, this table indicates that spontaneous pure creation only seems fully viable with the active presence of God. This is for several reasons.

Firstly, while it is fair to speculate about any mechanism in 'neutral circumstances', the presence of evidence which directly contradicts its existence must count as a significant challenge to such a theory. All the evidence we have indicates that Matter/Energy cannot do such things, both in terms of capability, plus the observations made by astronomers which go against such expectations, (ie. we do not see overlapping creation events, or large quantities of 'negative matter').

Secondly, even if we assume that it is possible, as one-off event based in Matter/Energy alone, most scenarios would place a great reliance on luck to guarantee our version of reality. There are convoluted ways in which this might be overcome, but they add significant extra requirements to the basic notion.

Thirdly, if spontaneity and pure creation only seem to be associated with the underlying stuff of Thought and a scenario accepts this stuff as real, then it removes *some* concerns but leaves others in place. Pure creation within a totally empty environment might become viable but as the first emergence of anything it would be reliant on luck. If pure creation occurred in the midst of an eternal existence then it is unlikely to have a major impact unless it 'swept the board', (as I have assumed that Intangible Thought *wouldn't* have unlimited power to shape the Universe).

Fourthly, pure creation seems highly unlikely to explain the source of the additional power needed to drive the 'sudden' acceleration in expansion of the Universe, because we do not see a burst of energy being cascaded from a point in space, *and* this acceleration effect is perceived to be a prolonged exercise not a 'flash in the pan'.

There are numerous other detailed reasons related to particular starting conditions.

Turning to the Materialist agenda, there is only **one** fully viable possibility and that is dependent on the presence of vast amounts of material contained in hidden dimensions of existence. (It is unproven, but there is no evidence to contradict the idea). Only three other ideas come close, (having one significant challenge), and one of these is dependent on the influence of Physical Beings with a potentially suicidal tendency!

Atheism has more scope because it can consider the potentially beneficial impacts of a second type of stuff on an eternal existence, as long as this doesn't extend to God. In directly comparable terms to the Materialist findings it now has 12 cells which are fully viable (6 underlying circumstances doubled-up to accommodate static and dynamic start points). There are a further 5 yellow cells (3 underlying conditions), which come close.

The additional circumstances are mainly achieved because spontaneity is now possible, however in a small number of cases the presence of Intangible Thought could exploit natural mechanisms (such as a Framework for Existence), to reshape physical matter without causing the self-destruction of the associated beings.

All of the other 31 fully viable solutions, (19 underlying circumstances), exploit the full power and awareness of a Creator God.

Cells which don't offer a reasonable chance of success

Row 10, (a movement/transfer of material within our 3 dimensions, due to natural currents/channels/tubes etc), is the only mechanism that **never** seems able to offer a realistic chance of generating our Universe. Row 6, (movements/transfers of material through the action of voids/vacuums), is barely any better, with the exception of God's involvement to perhaps create a large and perfect vacuum – however even this has a number of challenging issues which render it unlikely.

Row 5 dealing with the mechanism of Induction has a few unlikely possibilities (red) but is often held back by the need for luck in achieving our outcome. This is mainly because there is no apparent way for new configurations of existence to emerge, (as Induction works within the rules, not outside them). The ability to be spontaneous doesn't seem to change this in any reliable way, (ie. it still can't become part of an eternal cycle), including the concern that the fields would only have a limited stock of energy and therefore unable to produce many Big Bang events.

Within Materialism, the cells in column 'b' all suffer from the fact that a **perpetual** state of chaos would require a spontaneous or **ex**ternal act to suddenly break it out of its malaise and bring its material into some sort of configuration/order – even if it then explodes. This will effectively lower all ratings by 1 level.

Conversely, all cells in the 'static' columns c & e will not be functional unless they were made dynamic by a source of spontaneity so a 2nd type of stuff would raise the possibilities by 1 'notch'.

Within the Dualist sections, a source of spontaneity is available due to the presence of a 2nd type of underlying stuff, however the primary reasons why many cells do not seem viable relate to :

- The absence of explanation for the accelerating expansion of the universe.
- An inability to explain how the configuration of existence might change.
- A reliance on luck to achieve our outcome.
- An inability to explain the size and pattern of movement in our universe.

Key Points Arising From The Review

The table reflects our earlier considerations that the familiar Bang-Crunch theory based around an exploding singularity within our 3 dimensions not only has a number of serious challenges to its viability but that it now seems dependent on other types of stuff to make it a possibility (yellow), and even God to make it fully viable (blue)!

The issues primarily relate to concerns that:

- The eternal Bang-Crunch cycle would have ended, (due to a spontaneous or **ex**ternal influence), leaving us dependent on luck to achieve our outcome.
- A singularity could not be guaranteed to explode every time.
- It would be reliant on the unexplained process of Cosmic Inflation.
- It doesn't explain the source of power driving the accelerated expansion.

However if the basic mechanism is extended so that its explosion rips a hole through to other hidden dimensions, then some of these problems disappear because, (as discussed in 10.2.3), crossing dimensions can explain:-

a) **why the configuration of physical matter might change** – because it could force material to adapt to a different reality, (new Laws of Physics).

b) how the process of **Universe inflation** might arise – by increasing the power of the explosion at a moment when the material is free of all Laws of Physics.

c) How the power to drive the acceleration **could be delivered as a trickle to all parts of existence at the same time,** (rather than travelling billions of light years across vast distances in space.

The factor which would remain unresolved is that the cycle would still have ended, rendering it finite instead of infinite and therefore potentially reliant on luck.

One way to overcome this is for an unlimited cosmos to generate an infinite number of other Big Bang events (eg. singularities) elsewhere, producing every version of Universe somewhere in the totality of existence. (This could resolve similar difficulties within the concept of 'splitting of nothingness', cell 4d). An alternative is to suggest that a stable physical reality could **only** emerge in one form - our version of reality.

Whether or not you like these suggestions they are logical possibilities, although the issue which seems to prevent 4d being fully viable is the need for a spontaneous or **ex**ternal trigger to generate the split in nothingness, if there was initially only this stuff around - because no other factor would exist in the Materialist argument, (beyond the known scope and properties of Matter/Energy).

Moving onto the next narrow section of the table above dealing with Physical Beings - (7th column 'd'), this summary only shows one column from the underlying table which has the full 6 columns 'a to f' – the one dealing with scattered dynamic matter. This is because physical beings can only exist in the circumstances of column 'd'.

We know from the last chapter that physical beings have the opportunity to exist prior to the last Big Bang in the context of an eternal physical existence that is undergoing a cycle of change/transformation, yet they have their limitations in two critical respects.

Firstly, they are reliant on harnessing natural mechanisms to give them the power to change the Universe, and in many circumstances this would add little or no strategic benefit to an eternal cycle because these beings would only be able to accelerate the processes that nature was already deploying, (to provide something that suited them within their lifetimes).

Secondly, in the few circumstances where they might exert an influence to help generate a *new version of reality*, (a designer Universe), their limited knowledge and 'potential to make mistakes' could be catastrophic by ending the cycle prematurely - counteracting those potential benefits.

Yet if you combine these two points there is a more unsettling possibility – that a natural capability to end the eternal cycle could be deployed by nature anyway, making those mechanisms finite, and therefore dependent on luck to achieve our outcome.

We should also recognise that any attempt to implement a new designer configuration for the universe would almost certainly destroy any physical beings within it, and if the necessary outcome wasn't achieved those physical beings couldn't try again.

While there is comfort in knowing that we are here and that the outcome from such an experiment would have been successful from our viewpoint, it is still a mechanism that is dependent on luck. Therefore, to avoid those beings worsening the chances of achieving our Universe and to provide a more logical course of action that allows them to survive, it would be more realistic for them to *prevent* the next Big Bang occurring, in order to preserve their existence. This might also be achieved through a mechanism which falls within the Laws of Physics and is therefore more predictable.

Moving on again, although I realise that the idea of 'Intangible Beings' will be anathema to some people, the spirit of this exercise means that all possibilities should be considered for their ability to achieve the right outcome, before personal judgements are made about how credible they might be.

Like it or not, some processes of origin could be improved if intangible beings were able to implement a design, especially when they wouldn't necessarily be destroyed by a fundamental re-shaping of physical reality, so they could try again if things went wrong, (if the general circumstances permitted it).

Such beings would not be Gods but they *could* operate in environments that wouldn't support physical beings, (eg. an absence of physical matter). For instance if pure creation was building up a stockpile of intangible stuff that contained energy.

Their life spans might also be longer than physical beings. On this basis, they could be a source of spontaneity and a trigger where no other trigger was possible; but they would still be reliant on harnessing natural mechanisms.

Their greatest potential would probably lie in utilising a Framework for Existence, (row 12). If the civilization could insert new designer parameters into a Framework they would be applied instantly everywhere – either guaranteeing the outcome if they didn't make a mistake, or with the ability to try again later.

However there would still be a need to explain the accelerating expansion.

The final set of circumstances relate to the potential offered by God. At this point I feel that talk of 'percentage improvements' would be misleading. A Creator God could achieve anything, it's just a question of how we label things, and to make the table useful I have used God to demonstrate the maximum potential that each natural process could offer. It is only in Row 2 that his unlimited powers could be unleashed.

The significant exception to this is in Cell 2a which is shown as yellow because there are different ideas about what God is. In particular, there are different opinions about what God might be made of. If it is the same stuff as exists in our 3 dimensions, then an absence of everything would also mean an absence of God. However if God is made of different stuff beyond our 3D space, He might also influence 2a by creating the right circumstances for pure creation of physical matter to arise. Across these options our personal view of what God might be will dictate the possible outcomes.

General Points

The opportunities discussed above will only crystallise if the mechanisms can be shown to exist, and of course, as things stand, none can do so. We each have to make a personal judgement on what we believe to be real, however we can now do so with much greater confidence based on which mechanisms might succeed.

I will return to these questions in the conclusions section at the end of the final chapter.

Across the 4 exercises to assess the 70 different process combinations, a small number of key factors kept featuring as pivotal issues and need to be borne in mind:

- In several circumstances we couldn't expect natural circumstances to trigger the appropriate mechanism when needed. This generally applies to columns 'a & d'

- While Row 4 envisages that 'Nothingness' could be split into balanced opposites, it is by no means certain that those opposites could only take one form rather than producing different versions of positive and negative existence each time the process was run.

  This is not just a case of antimatter vs negative energy, it relates to the configuration of existence and the Laws of Physics that might apply. There could be many different types of outcome.

- Suggestions that mechanisms intended to preserve the structure and integrity of existence, (eg. a circular chain of reality), led to a Big Bang event must consider **how** a significant imbalance could arise in these circumstances – given that the mechanisms would be there to **prevent** an imbalance from occurring.

- A rip/tear in the fabric of existence, or the ability to bend space in order to create a 'wormhole' through which material can be transferred, must assume that there is a 'fabric of space' that can be manipulated. I have assumed that such a fabric could exist in column 'a' to help define our 3 dimensions and distinguish them from other hidden dimensions. I do **not** believe that this should automatically extend to the existence of a Framework for Existence intended to regulate the shape of anything that exists within the dimensions.

As mentioned before, the results I have presented above reflect my attempt at this exercise based on certain assumptions. You may make different assumptions or have other beliefs about the likelihood of each scenario.

However, whether you use your view or mine, once this analysis is available we can then move-on to produce a map of the viable/possible routes to our existence.

12.2   <u>Tracing the Paths of Ultimate Origin</u>

The analysis conducted in chapters 10 & 11, plus the summary table above, provide us with the viable routes which might have brought us to our present reality and that gives us an opportunity to present them in pictorial form – as a map.

In order to make sense of the full map I will briefly summarise the different factors that we have encountered so you can trace them onto the overall diagram.

The concept of the Bang-Crunch cycle should be familiar to you and could either be represented as a mechanical process or a more philosophical concept :

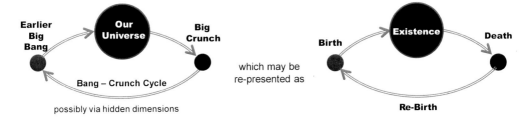

On the left hand side we see that an earlier Big Bang leads to our inflated universe, which then collapses again into a singularity through a 'Big Crunch', before the cycle repeats itself. This has direct parallels with Saṃsāra - the cycle of Birth, Life, Death, and Rebirth which the major 'Indian religions' believe in, (although some believe that the eternal cycle is driven by the gods rather than it just being a factor in nature). This is what I represent in the diagram on the right – a generic version of the cycle.

The generic version is also useful because it is applicable to any cyclical activity which re-uses exactly the same material to generate new Universes through a process of transformation... although it is easy to think in terms of Bang-Crunch.

As we have seen, the key scientific discovery which broke the Bang-Crunch cycle is that the expansion of our universe is accelerating, which means that it cannot collapse again, (if that had indeed been the previous pattern).

It also means that if there had been an eternal sequence up to the last Big Bang, something spontaneous or **ex**ternal must have occurred to end it.

> This would be true even if our existence had infinite size and contained
> an infinite number of other Universes, because they would **not** have
> exerted sufficient influence on our Universe to break the cycle sooner.

So in this respect an **ex**ternal factor must either be completely separate to our physical existence, or be out of sync with it.

Thought is the main factor in our universe which seems capable of random or spontaneous behaviour, or might act in as an **ex**ternal influence (because it was 'out of sync' with the rest of the natural environment). The use of Thought may either come from physical or intangible beings, (using one or other type of Thought), or from God. This allows our diagram to evolve as follows:

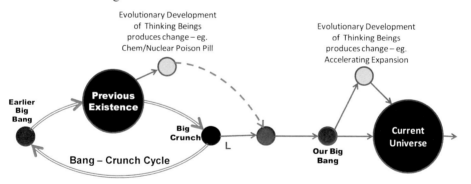

If the eternal cycle was broken at the last Big Bang, (13.77 billion years ago), then the acceleration which began 5 billion years ago could represent a second instance of spontaneous or external influences. In the diagram above we see that the bang crunch cycle may have produced life in either a prior universe or the current Universe which either led to a change of configuration, or to generate a larger bang than normal, which therefore prevented it from collapsing.

If we exclude the possibility of Thought in a prior Universe some other spontaneous or external change must have occurred in Matter/Energy – which would require luck rather than design to account for our version of reality, (hence the presence of the 'L').

However it's possible that the eternal cycle didn't end at the last Big Bang event. It may be that the *sudden acceleration* was the factor that broke the eternal cycle – in which case there may only be **one** instance of this spontaneous or **ex**ternal influence.

The acceleration in expansion which began 5 billion years ago definitely falls within the lifespan of our Universe, (having occurred well after the last Big Bang). It also cannot have been a natural feature of the eternal Bang-Crunch cycle as it would have ended that cycle well before it could explore every permutation of reality – so it must represent something new, which arose just before the emergence of our *solar system*.

As it has taken 4.5 billion years for sophisticated life to emerge on our planet the prior 8.77 billion years seems more than adequate to allow an earlier civilization to evolve and become even more advanced than us. The only alternative sources for an end to Bang-Crunch are raw spontaneity or God, (the only other perceived **ex**ternal factor).

Yet Bang-Crunch isn't the only explanation on offer. As this had been the principal Materialist explanation of origin and it was now broken, it was inevitable that other eternal mechanisms would be proposed that were unaffected by the acceleration. The most prominent of these new theories, (that membranes/curtains of energy are 'hanging' in space or within other dimensions beyond our view), illustrates this point. It is clever in making the unceasing expansion of the Universe a necessary part of cycle, and it also differs from Bang-Crunch by suggesting that each Big Bang event will use **different** material, from an infinite stockpile.

By essentially arguing that the eternal cycle remains unbroken there may **not** be a need to look for a **spontaneous or external factor**, however the acceleration effect is still real and needs to be explained. At headline level, the scientific belief that this effect began well after the last Big Bang, suggests that this was a new start point, which would again mean that a spontaneous or **ex**ternal factor may be required. To counter this Materialists have argued that **the acceleration marked a tipping point in an inevitable process**. More specifically, it was caused by an effect which had been there ever since the Big Bang but which either needed time to grow to sufficient strength, or for influences that were restraining it to diminish.

In overview we don't know which of these polarised views is correct. They are all speculative but one suggests a beginning whereas the others suggest the next step in an eternal process. However, as with the Big Bang itself, Materialism is trying to defend the less instinctive view by saying that the start point is not a true beginning.

Our diagram needs to reflect the core idea of a simplified cycle that is also immune to spontaneous/external influences. As the 'curtains of energy' notion says that each universe will be spread out but never disappear, there is no need for a specific 'death' scenario – just a succession of new Universes. This can be illustrated as follows :-

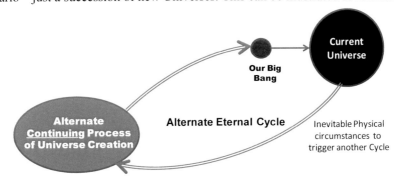

Of course, such a mechanism would need to emerge in some way if it is entirely a product of Matter/Energy, which means that we have to link it back to ideas of prime origin. This is where we must remember that eternal existence is not the only option.

Logic tells us that there are just two core possibilities for prime origin:- pure creation, or eternal existence. In addition to these we can layer-on other factors to give us a more detailed set of scenarios:- by indicating whether or not there was a start to Time; and whether Matter/Energy had a different start point to a 2nd type of underlying substance, (the stuff of Thought). The diagram below shows the options which exclude Thought.

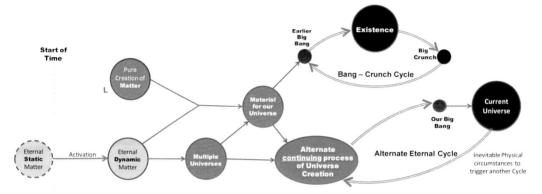

We see that *eternal* physical Matter/Energy could exist outside Time, but it would be entirely **static**, yet it would have to become active before the Bang-Crunch cycle could begin. The activation of Time, (an injection of movement), would be necessary.

Alternatively, if all Matter/Energy had a true beginning and spontaneously emerged from pure creation this could only be *within* active Time as it would be a dynamic event, (requiring a sequence), and this would lead directly to our Universe, even if there were other instances of pure creation which formed other universes. If existence started from a 'blank sheet' we have to assume that the **capability** for spontaneity and pure creation would always be there, even if nothing existed beforehand.

Within these scenarios are the possibilities that eternal existence occupied other hidden dimensions, or that it may have existed as the zero form of something, (Nothingness). With enough time, the different factors may also have been able to develop a 'fabric of space' or even a Framework for Existence. This reflects the position for Matter/Energy

The equivalent ideas for a 2nd type of stuff that underpins reality looks like this:-

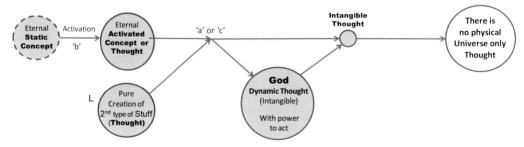

We again see the options of eternal existence or spontaneous pure creation, (arising from absolutely nothing), but this time for Thought.

186

If the stuff of Thought existed before the activation of Time then it would again be entirely static and may therefore represent a series of concepts that did not interact. The starting of Time would make this stuff dynamic, and would also begin the process of true interacting and creative Thought.

If God exists then He will be made of something, and from the two types of stuff available in the overall diagram I have chosen to portray God as being based in the Intangible stuff of Thought. As the ultimate example of an Intangible Being, and a potential shaper of our awareness and our minds, it is logical to show God as a source of existence based on Idealist principles.

Yet we should also try to show how Thought can work with Physical Matter.

The Dualist position says that both types of stuff exist, and there are various conceptual ways to bring this about. Aside from the issue of whether either of them is eternal, there is the question of 'Did they arise independently, or did one type of stuff generate the other type?' All of the different permutations are conceivable.

If either of them existed first, then that stuff could potentially generate the right circumstances for pure creation. It may also be able to influence the early development of the other type of stuff. Without any evidence to indicate which of these options may actually be possible, all arguments appear to have equal viability.

However, to separately show each of these 'potential sequences' would make the diagram very complex, so I have found the presentation is greatly simplified by showing that the flow of influence could operate in either direction with a double headed arrow, as shown in the diagram on the left below.

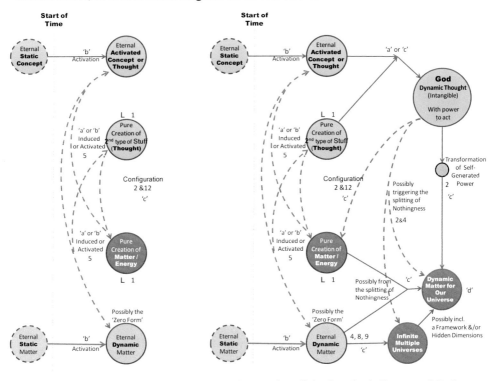

Basic start points   Start points linked to the influence of God

187

The diagram on the right (above) accommodates the possible actions of God. Logically this would either be to convert part of Himself into physical reality, (injecting His power as raw energy into our environment per the solid orange line), or configuring the existing material, (dotted orange lines), so that it was ready to generate our universe through a subsequent process of evolution.

However the options for God extend beyond this because He could apply His power at different stages within a natural process, or indeed avoid evolution to generate our Universe directly. This is shown in the section of diagram below.

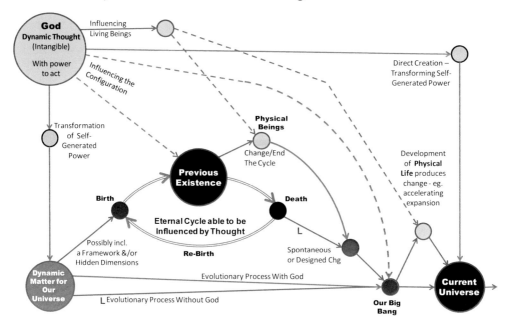

If we follow the solid orange line across the top and then down on the right, we see the traditional concept of the creation story where God directly fashions our universe, avoiding a cyclical Big Bang process - going straight to a fully formed Earth.

Alternatively, if we follow the dotted lines, we see where God may provide the spontaneous/**ex**ternal factor which broke the Bang-Crunch cycle to create our Universe. The orange lines represent his direct influence over physical Matter/Energy while the blue lines indicate His influence over living beings, (Thought).

These are ways to represent religious ideas which have not been disproven, however in these diagrams all paths are possible, so God's influence isn't necessary to each theory, (ie. route through the map).

To complete the rectangle, if we travel down the left hand side and then across the bottom, we see the creation event discussed earlier that sets-up the basic material of the universe (grey 'Dynamic Matter' on the left), before allowing it to follow a natural course of events without any cyclical activity. If evolution follows an inevitable course then the start point set-up by God would pre-determine the outcome, but without His influence or a cyclical mechanism with which to experiment, the emergence of our specific configuration of reality would be down to luck.

While there are various aspects of the Bang-Crunch theory which **require** a spontaneous or external influence, this is not a logical requirement of other cyclical activity, and to make this point I have **not** shown God's influence on the other cyclical mechanisms, (although this is a possibility even if it isn't necessary). It also serves to prevent the diagram becoming over-cluttered.

With this one exception all of these elements can now be assembled into an overarching diagram that reflects the full range of possibilities, (as shown in 12.3 below), and hopefully you will now be able to recognise the different components and what they represent.

Before I reveal this however I want to summarise what we have done and place a strategic context around the overall set of possibilities.

12.3 **A Map of the Paths to Our Reality**

Different styles of diagram appeal to different people however the 'map' that I present below is the simplest that I have been able to achieve which covers virtually all of options. At first glance it still seems quite 'busy' but hopefully with your knowledge of what all the different components represent, it will be relatively easy to follow.

The map works from left to right with a start point of either eternal existence or pure creation for both types of stuff that may exist, (4 places to start).

Each of the paths marked by the arrows indicates a process which changes the starting circumstances into something else via one of the mechanisms we discussed in the earlier chapters, (referenced using the row numbers from the tables in 12.1).

Through this, I map out the different sequences of events that may have occurred.

This is different to the analysis I conducted in 12.1, which was purely to test *viability*. Here we are looking at sequences: for things that may interact at different stages.

As an early example, you might think that the 4 starting points would become 8 if we also said that each of them could begin in a static or dynamic state – but things are far more complex than that. I found 255 significantly different permutations because the 8 basic conditions can be combined in different ways. For instance if Pure Creation produced entirely static physical Matter/Energy, this could be combined with eternal static Matter/Energy and eternal dynamic Thought. This is just a combination of 3.

If you consider that these 8 starting conditions could be combined in groups of 2, 3, 4, 5, 6, 7 or 8 you can see how 255 options might arise. If we increased the number of base situations by adding extra factors, (such as whether the different types of stuff were scattered about or concentrated in one place; or by including a full set of options for Time), the numbers grow very dramatically, becoming unworkable.

The benefits of a diagram start to become apparent – even if there are some modest short-cuts. Those short cuts are to aid understanding but not to remove quality.

As an example, if there are a significant number of permutations for dynamic vs static substances, we may validly reduce the options if we know that only 2 outcomes can ultimately emerge in our current reality - dynamic Matter/Energy with or without the dynamic stuff of Thought. **Outcomes can keep us focussed on what is relevant**, and in this case, dynamic things can be assumed to always activate static things where there is an interaction.

189

There are other significant ways in which we can simplify the diagram, so it's worth noting these broad factors:-

- Some combinations are effectively duplications of others in the list.

- It is pointless to show some possibilities, (eg. if **spontaneously created static matter** added to the stock of **eternal** static matter - even if these are possible occurrences through the end-to-end process).

- Some scenarios only became viable in specific types of base situation, (eg. dynamic things can't exist without Time).

However, we should also be cautious about removing options just because we may not like them. For instance, most people would envisage that the spontaneous creation of matter would result in a dynamic explosion of material rather than a static blob – yet both are conceptually possible even if we feel that one is more likely than the other.

Using this type of logic the diagram must show the high level stages in each logical process, from starting conditions through the actions of mechanisms to the end outcome.

In doing this I found that each path between the circles represents one of 4 basic types of influence/activity that represent an interplay between Matter and Thought:

a) The possible need for one elementary factor to create the other,

b) The need to introduce movement where the original manifestation of Matter or Thought had been static,

c) The need to configure or shape physical reality,

d) The possible need to aggregate physical matter into a singularity if it had originally existed as scattered material across space.

I have used these letter codes as labels on the map which can help us to think about what is happening at each stage in the process.

At various points in the diagram I also indicate where luck would have to play a part, (using an 'L'), and as a link back to the 12 mechanisms in the analysis tables earlier, the numbers indicate the mechanisms which might be deployed at each stage.

As a way to become familiar with the diagram and to see its usefulness it can be interesting to trace a path from left to right while avoiding the need for luck.

You may be surprised at how few of those routes turn out to be possible.

If you then try to avoid Thought it becomes very hard indeed – and this is in keeping with the viability assessments per the table at the start of this chapter.

Inevitably there are limitations to what this diagram can show but it does cover most permutations with an element of 'shorthand' as described earlier. In terms of detail it needs to be interpreted alongside the viability tables in the previous section.

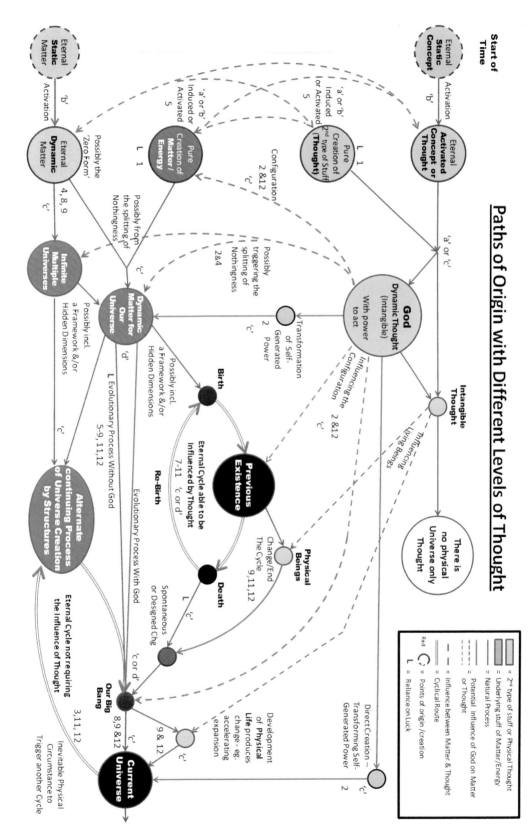

Paths of Origin with Different Levels of Thought

191

Review and Conclusions

Our earliest ancestors were born into the middle of an existence that they simply didn't understand, either in terms of scale or complexity. To an extent we share the same fate because although knowledge is passed-on from generation to generation, we still cannot answer many fundamental questions such as: how big the Universe is; how many layers of reality there might be; and how many types of stuff underpin it.

All three of these questions try to quantify the scale of existence by seeking different forms of boundary around it, and yet we still don't have an answer to any of them.

While all of these questions are valid and necessary, the research undertaken so far has simply indicated that all of the answers lie beyond what we can see directly, therefore we will be reliant on more advanced technology & understanding to uncover the truth.

Equally we face a dilemma in knowing when we may have reached one of these boundaries because none of these issues will have a clear marker to show where the limits are. All we can do is to look for gaps in our understanding, and determine if any evidence pushes the suspected limits further out.

As part of this process of discovery, the role of science has been to establish facts while philosophy has structured our speculation about the unknown. When the disciplines of science and philosophy work together we should find that facts serve to narrow the range of credible speculation.

We can genuinely celebrate the achievements of science in identifying some of the rules by which nature seems to operate. These allow us to make accurate **predictions** of how Matter/Energy behaves - demonstrating that our understanding of the related subjects is complete. Although those rigid rules cannot anticipate human behaviour they are increasingly being used to explain natural events in the *past*. In this respect the predictions try to anticipate discoveries that we might make about the origin of our world and our Universe, acting as a yardstick by which we can judge each theory.

> Until a couple of decades ago it was accepted as a virtual fact that while the Universe was 'exploding' outwards, its expansion would be slowed and then reversed by the force of gravity... but in reality the expansion is accelerating.

Where future events or new evidence do **not** coincide with predictions then we have to go back to the drawing board. In broad terms, scientists will either adapt a theory which has already brought us close to full understanding, or scrap the theory and start again because the principles being revealed by new evidence contradict the principles which had previously been suggested. However in truth, people can be very reluctant to change ideas which have served them well for many years. That is a feature of both science and religion – we are all human after all.

Through the course of this book I have tried to show you the evidence which is significant to the nature and origin of existence, either because it reinforces our understanding or because it directly challenges established principles.

While it is perfectly reasonable to be sceptical about dramatic changes in thinking, and to continue searching for answers using established foundations of belief, there comes a point when alternate ideas have to be taken seriously and not dismissed out of hand. That doesn't mean that the new ideas are automatically correct but if they fit the available facts, any new perspectives might help to unlock a stubborn mystery.

In broad terms, the findings which have been of interest fall into certain categories:-

a) those which revise our thinking about the *limits* of existence.

b) those which shape our speculation about the *structure* of existence.

c) those factors which mark a true beginning, or *a point of fundamental change*

d) those which influence our thinking about the *mechanisms* of origin.

e) those which shape our ideas about the *outputs* from the process of origin, (our current reality)

The missing factors from this list concern two things:

f) Ideas about the *starting conditions* for any process of origin, and these generally seem to be covered by philosophical logic rather than any evidence.

g) The reality of *Life and Thought* which scientific laws cannot predict, but which may have influenced the course of events.

As these factors all form part of a single story then it seems likely that their influences will intertwine, however I need some structure for this summary and the later conclusions, so I will base them around these seven categories.

The Limits of Existence

In terms of the size of the Universe, our increasingly powerful telescopes reveal more and more galaxies that are further and further away. We have not reached a limit, however more importantly, the distances being covered (93+ billion light years) greatly exceed anything that could be achieved by travelling at the speed of light since the start of our Universe 13.77 billion years ago, (as currently estimated). Put another way, the idea of an exploding singularity only works if it can break the Laws of Physics.

In addition, the results of the WMAP survey indicate that space is not curved and may therefore be infinite... and if space is 'something' which can be occupied then no known physical process could put it in place. Unfortunately there is little to add to these basic findings about size until technology develops further.

With regard to the number of layers in existence, scientists believe that there is a fundamental difference between the 'higher' level of reality that we occupy, (atoms or bigger), and the reality which exists deeper within existence, (smaller than atoms). This is because fundamentally different rules, (Laws of Physics), seem to apply, and yet they must be part of a single linked reality.

Interestingly, this is also the boundary between what we can see directly in real time, (even if that is using the most powerful microscopes), and what we can only infer 'after the event'. While it is the hope of some scientists that deeper truths can be revealed which will unify these layers within a single set of rules, that is not the case at present – and the differences between those layers are quite profound. Once again, we are dependent on more sophisticated technology to reveal more.

So our attention turns to the question of how many fundamental types of substance underpin existence? The standard answer from science for many years has been 'One', Matter/Energy, but a growing list of discoveries suggest that there may be others.

Certain philosophies have, for a long time, suggested that Thought may be something fundamentally different to Matter/Energy and I have sought to provide a scientific interpretation of why that may be, through the presence of spontaneity or randomness. If these capabilities exist then there is a strong case for another type of substance.

However the possibility of other types of stuff has also been raised as a serious possibility to answer major factors in cosmology.

'Dark Matter', (which may actually be transparent), has been proposed as a source of gravitational effects which cannot be explained by the physical objects we can see (stars, planets, gas clouds and galaxies), and having proposed this concept science **has** subsequently found direct evidence for this invisible material in the way that light is bent when it passes through Dark Matter. Yet we still don't know what this stuff is.

On the back of this success, science has also speculated about 'Dark Energy' as a way to power the accelerating expansion of the Universe, (the reverse effect we would expect from gravity), however as yet there is no specific evidence for such stuff and it imposes additional requirements on the logic of existence as we will see when we consider the points of fundamental change, later.

According to NASA calculations, the amount of Dark Energy that would be needed in the Universe to generate an even acceleration everywhere is vast, (approximately 15 times the amount of physical material we can see in all the galaxies of the Universe), and on this scale it should be a noticeable presence across the Universe, probably including our solar system, and yet we have not identified it. On the other hand it could be that we **have** encountered it but not recognised it for what it is.

An idea along these lines may be one way to explain the results of the Dual Slit and Quantum Eraser experiments, (discussed in Chapter 6), which, under certain conditions, cause photons & electrons to form a spread-out interference pattern instead of a block image. In these experiments we perceive **particles** to be issued from a laser which are then always recorded as a **dot** on screen after passing down the apparatus. The difficulty is in explaining both the spread and the fragmentation of the image to form an interference pattern, (which is commonly associated with a spreading wave).

The only way to explain these 'polar opposite' effects within a single type of underlying material is to argue that this stuff changes form, and 3 ways have essentially been proposed:

- Particles turn into spreading waves and then back into a dot, (for some unknown reason choosing to materialise at one point along that spreading wave).

- Things that we perceive to be particles, (photons & electrons), are actually intangible 'fields of influence' that spread out as waves, but the waves then have to crystallise into a single dot at the end – not a myriad of faint dots across the interference pattern.

- The underlying stuff has no form until it is required to crystallise into something appropriate to the circumstances, (it leaves the laser without form but crystallises into a dot when it is detected – the position of that dot relating to whether there is certainty in the path taken).

These explanations are hard to accept because they overturn our natural perceptions of what existence is, yet people do this because their belief in a single substance is so strong. However even if we accept these convoluted explanations there are some significant problems with the ideas.

For a start, they struggle to explain all of the facts. As an example, the same results are apparently produced when the experiments are re-run with carbon bucky balls which definitely have a firm shape and do not change into a wave en-route. To say that those objects have no shape/form when they are observed, would extend the different rules being proposed by Quantum Mechanics for the deepest layers of existence, beyond atoms and directly into our level of reality. In effect this is the realm of full Idealism.

It may be true but it is far from proven, and many will believe the evidence of their experience of life to say that it isn't correct.

We should also consider what the symptoms tell us about underlying circumstances. In simple terms there are two types of wave which should not be confused.

There are wavy lines that travel in one direction, (typically associated with beams of light), and there are spreading waves which are associated with ripples on a pond.

I suggested that spreading waves are **only** produced by a pool of 'stuff' when objects pass through it. We can liken this to a line of boats causing ripples on a lake/ocean, or a train generating sound waves as it travels through a pool of air. *Spreading* waves always arise in the *pool*, **not** by transforming the things which travel through it. The hidden pool has parallels with the deBroglie-Bohm hypothesis[33].

We can also draw the analogy that a line of ships (particles in a beam of light) would pursue an up and down wavy course over the waves, but where there is interference in the pool, those ships might be caught by the enlarged troughs of the waves generated by an interference pattern, causing them to change course. By following those channels the particles would inevitably create a negative image of the interference pattern when they hit the screen. In other words, the bright areas of the screen would be where the troughs were.

At first glance talk of a hidden pool of other stuff may seem like far-fetched wishful thinking and yet we can draw parallels with the pool of Dark Energy whose perceived scale must surely be everywhere... which we can't detect either.

So could a hidden pool of other stuff be a plausible and simple explanation for the Dual Slit results which have puzzled scientists for so long? It would allow our long held beliefs about the nature of particles to remain intact, and would squarely fall within the realm of Materialism if this hidden pool did represent the same stuff as Matter/Energy. However, equally, it might represent very different stuff instead.

The main experimental findings that would **not** be explained by a hidden pool are :-

i.   why interference patterns disappear when monitoring equipment is used to determine what is happening at the slots in the Double Slit experiments;

ii.   why D0 results are the same as the other detectors, without resorting to time travel in the Quantum Eraser experiments; and

iii.   the faster than light communications and sense of mutual awareness displayed by paired particles in the Gisin and Bell's Theorem experiments.

While the answer to 'i' may be quite simple, (ie. the equipment interferes with the hidden pool, not the path of the particles), points 'ii & iii' are difficult and do seem to indicate other capabilities which may be related to data/information.

In overview therefore, while Materialism might still be proven correct, the grounds for believing in other types of stuff have gained credibility for four main reasons :

1.   There is both physical evidence and a logical need for capabilities that go beyond those attributed to Matter/Energy,

2.   Other types of stuff would allow our ideas about proven aspects of our reality to remain unchanged.

3. Explanations about unresolved factors in existence can be greatly simplified.
4. The potential evidence for other underlying substances comes from different disciplines which are beginning to complement each other, (cosmology & QM)

While Determinists and most Materialists will deny that spontaneity and randomness exist at all, it becomes significant that these characteristics are required by many theories of existence to explain fundamental change, such as

the same singularity adopting different configurations of reality each time it explodes; or the ending of an eternal sequence; or the starting of new things.

As Thought is the only thing we are aware of that may be capable of spontaneous or random acts, the ***underlying stuff*** of Thought is potentially linked to those situations... if it exists, (either as a separate substance, or new properties within deeper levels of existence). If it doesn't exist then those theories may need to be changed.

Put another way, where cause & effect within Matter/Energy fails to explain a situation the only other logical way to explain fundamental change is via an **ex**ternal influence or a spontaneous/random effect.

Theories like the exploding singularity have just assumed that a singularity would enable fundamental change to occur, without any explanation, and yet a singularity must be part of an ***unchanging set of rules*** which would allow an eternal Bang-Crunch or other cycle to take place. As another example, genetic mutations cannot appear by Matter/Energy 'making a mistake', as has been claimed. Cause & Effect doesn't allow for mistakes, there can only be randomness, spontaneity, or **ex**ternal causes.

In summary, the characteristics which point to other types of underlying substance that may underpin reality, (point 1 above), come from :-

- The need for either spontaneity/randomness or **ex**ternal factors to explain change
- Dual Slit Experiments – spreading wave effect requiring a hidden pool, and suggestions that particles may be 'aware' of their circumstances.
- The findings that paired particles have a sense of mutual awareness and instant communications.
- The creative characteristics of Thought, and our belief in Free Will.
- Dark Energy as an additional source of power **beyond** the material which emerged from the Big Bang.

At its core, speculation about other stuff is a way of saying that certain characteristics and capabilities are real, **and** those effects are quite distinct/separate from the normal workings of physical Matter/Energy. It may be that these effects are drawn from much deeper characteristics of physical matter which we don't normally see, (a Materialist view), but the separation would still be real, so the characteristics attributed to other types of underlying substance would, for the most part, still be valid. It is equally possible that other stuff does truly exist, (the Dualist and Pluralist perspectives). Conceptually I find it easier to consider these factors as separate substances.

Whether it is intangible or not, 'other stuff' could potentially adopt a variety of forms with different levels of sophistication, (say crude, middling, and high end), in much the same way as we see Matter/Energy taking different forms as it evolved. At different stages in its development such stuff may therefore be able to add different layers of capability to Matter/Energy, whether that is a hidden source of power, spontaneity/ randomness, the ability to change the Laws of Physics, provide a basic sense of awareness/direction in evolution, or even full-blown Thought.

The Structure of Existence

Physical reality has structure. We can all see that. The question is why?
Four main issues/questions surround and drive this debate.

a) Does space have to exist before it can be occupied? Put another way, is there such a thing as the fabric of space?

b) If the underlying stuff of physical matter resembles raw energy, how could something so fluid become hard and structured?

c) How can the Universe operate to a standard set of rules when scientists believe that there could be a myriad of possible configurations arising naturally?

d) Could additional dimensions of existence explain some of the findings that we struggle to understand?

On top of these we can layer-on the human demand that we explain how our version of reality emerged without resorting to luck.

The first issue is whether existence extends to those places which are empty but which could be occupied by something, (eg. space beyond the Universe)? There would be nothing there, but the fact that something could exist in that location means that it is a place, and we can potentially identify that place in relation to where we are. So is it part of existence?

That is a personal choice, and I suspect that people may view it in different ways depending on the question being asked.

In general terms, if empty space **is** part of existence and it extends to infinity then it cannot have been created by a localised explosive event, (regardless of how powerful that explosion might be), because all physical events that we know of are finite.

On this basis, if existence is infinite it is almost certain to have been that way for all time, (without a beginning). In other words, if we envisage a *beginning* to existence it will probably be finite both in terms of size and time/age.

The hint of doubt comes from one highly speculative possibility which is that 'spontaneous creation' might occur on an infinite scale, thereby providing a beginning to an infinite existence. Yet this would imply that the *potential to be infinite* was already present... ie. that existence was already infinite but without physical content. A beginning from pure creation would also have to point to either a spontaneous effect or an external influence like God or a parallel universe.

> To make the same point in a different way, if God put an infinite physical reality in place it would probably mean that God was already infinite – and a God that exists, is existence.

However if space is not infinite then we have to form a concept of what factors may limit it. The two main options are that :-

- there is a fabric to space and that existence only extends as far as the fabric, which may either be without beginning but limited in size, or may be created/ generated slightly ahead of any material that was moving into the void.

- existence can only be counted where 'stuff' exists whether that is physical Matter/Energy or something else, and if that stuff emerged from an explosion, existence would only be as big as the spread of that material.

If there was a 'fabric of space' the possibility exists that this fabric could be distorted by the presence of physical material. Gravity, as a force of attraction, was originally conceived as possibly curving the fabric of space into a ball in order to limit the size of the Universe – and we will remind ourselves how this could be translated into another type of force in the 'Mechanisms' section below.

Yet staying with the main theme, the findings from the WMAP survey indicated that space is **not** curved and may therefore be infinite, so would it also be inevitable that existence must be eternal too? I don't think this is a necessary conclusion, for 2 further reasons:-

1. We cannot be sure that existence does run to infinity. For instance, it could be argued that space is straight within our line of sight but curved beyond that – limiting the Universe; or alternatively the fabric of space follows straight lines up to a certain point and then stops, (like the edge of a flat earth).

2. We might distinguish between *physical* existence and the *totality* of existence if we say that the totality is infinite but that our Universe began in an explosion and the physical content is therefore finite.

Perhaps one way to find personal answers to some of these points is to think the opposite way – is there anything to stop physical matter from extending to infinity?

Moving on, in relation to issue 'b' above, the two main ideas that I am aware of which try to explain how structure might emerge out of fluid energy, suggest that :-

x) Energy is associated with tiny forces that may push on each other or bend round to press on themselves, thereby forming containers for energy which in turn generates 'substance', (the core idea within String Theory and M-Theory).

y) There is a Framework for Existence which imposes rules and structure on Matter/Energy, (possibly linked to ideas about a fabric of space), and would explain the uniformity we see in our Universe, (which appears to use the same Laws of Physics everywhere - ie. the same configuration of existence).

Such a Framework could also be used to radically change the Universe if a way of generating new settings could be found, (eg living beings).

Option 'x' still leaves open many questions about how those minute forces might be formed; where they start and stop; and how their strength, shape, and other necessary characteristics may be determined. On this basis it may be that both options 'x & y' are needed to make sense of what is happening. Sadly we don't know if either option is close to the truth. While a Framework for Existence might also explain of how the Universe standardised on one version of reality, we have also seen two other suggestions for this uniformity which could also apply to option 'x':-

i. that there may be a self balancing mechanism within the Universe which effectively forces different versions of reality to compete on a 'survival of the fittest' basis.

ii. that factors within existence mean that only one version of reality is possible.

Finally we considered the possibilities for other hidden dimensions of existence, although there is no physical evidence to support this notion.

Other dimensions proved to be remarkably useful in our analysis because they may not only explain Time and movement, but could also provide hidden stores of material. That material would almost certainly have a different configuration to our own 3D environment and would therefore provide a way to generate huge explosive events travelling at speeds which might greatly exceed the speed of light, if that material crossed into our 3D space without the Laws of Physics, (as discussed in Ch 7)

Factors which Mark a Moment of Fundamental Change

The significance of change is that it can only occur in two fundamental ways:-

- a)   As the next step in an inevitable sequence, or
- b)   As something genuinely new which would require spontaneity or randomness.

As Matter/Energy is only deemed to be capable of operating within causality, the only explanations provided by traditional materialist/determinist thinking have been via option 'a'. Unfortunately we have also seen many circumstances when the mechanism for change has been ignored by scientists and just **assumed** to be possible.

> For instance, as part of the Bang-Crunch theory it has just been assumed that an exploding singularity would bring a different configuration of reality with each new universe – which ignores the logic that any eternal cycle would need to preserve some of the Laws of Physics in order to allow the cycle to continue, plus the fact that exactly the same material would be involved on each occasion – which would mean that there is no apparent cause for fundamental change.

> This is particularly true when we look at the detail of this proposed mechanism and have doubts that a singularity would explode. Supermassive black holes, (the only objects which vaguely resemble them), don't explode, but vent their energy instead. Where the mechanisms of change are the key to any proposed eternal cycle, they cannot just be skated-over and assumed to work.

Causes of **fundamental** change **have to be** found in randomness or **ex**ternal factors.

We have seen that there is a fierce debate between those who believe that spontaneity and randomness might be real and those who do not, and I have hopefully demonstrated how difficult it is to prove a case either way when looking at events. However **if** such capabilities do exist, then they cannot come from Matter/Energy – at least not in the forms that we experience Matter/Energy at our level of existence.

Most examples of spontaneity/randomness are blurred by 'circumstances' plus the intricacy of human thought, so it is mainly when we come to issues of fundamental change within the physical environment as part of explanations for origin, that the need for spontaneity/randomness becomes compelling/unavoidable in some scenarios.

This is why points of fundamental change hold such significance; they draw us towards issues of creation; the different types of stuff that might exist; and the mechanisms that might be involved in the process of origin.

The beginning of our Universe must represent a point of fundamental change, whether it is the next step in an eternal sequence, or a moment of pure creation. The origin of Life is another major point of change, and more recently the discovery that the expansion of our Universe began to accelerate 5 billion years ago is another point of great significance.

However there are more subtle issues in the story of origin. One of the most profound comes from the realisation that the **ending** of an eternal cycle can only arise from a spontaneous or **ex**ternal event – and there are very few things we can envisage which are **ex**ternal to the physical Universe we inhabit, (ie. beyond the totality of everything physical). Other universes, parallel dimensions, a 2nd type of stuff, and God are the four main ones which spring to mind.

Philosophical logic also leads us to 2 other key factors which frame our concepts about the 'starting conditions' for existence :

- Whether or not there was a beginning and therefore a need for creation and spontaneity, plus
- Whether or not the original conditions were entirely static, or dynamic which would determine if there was a need to start Time.

If you therefore believe in any moment of true change, (whether a start or a stop), you will have to accept the need for capabilities beyond our Matter/Energy, whereas if you don't accept that there is any true change you must show how things can continue.

Mechanisms of Origin and the Key Factors that they Have to Explain

Various factors have led scientists to believe that there was a major event 13.77 billion years ago which we now refer-to as the Big Bang, however we don't know if that event was a moment of pure creation or the latest transformation of eternal material.

Our considerations about the origin of our universe have been framed like a standard process, with 'inputs', a 'mechanism', and a specific outcome.

We don't know which mechanism was active in the Big Bang as there is no evidence to indicate any of the proposed options. I identified 12 'mechanisms' that might be able to generate an explosive event on the scale of a Big Bang and many of these had permutations on a theme. We considered each of them in the context of 6 generic starting conditions, and discovered that the vast majority had one or more significant problems with their credibility. For instance, the familiar concept of a singularity can have different forms, representing eternal existence, pure creation, or a transfer from other places/dimensions. There was also doubt over the Bang-Crunch theory because the accelerating expansion means that it has either ended or was never correct.

Each of the 12 mechanisms could potentially be enhanced by different types of Thought and we identified 4 types of being associated with different types of Thought:

Physical Beings, Pluralistic Beings, Intangible Beings, and a Creator God.

There is no doubt that Physical beings are possible **because we are here**, and in a scenario where eternal existence cycles through a series of different universes with different configurations, an **advanced civilization of physical beings is a distinct possibility**, and arguably inevitable. Pluralistic & Intangible Beings plus deities, (as commonly perceived), would require other types of substance to exist with different characteristics to Matter/Energy – and we have seen the logic to justify such thinking.

The notion that God is allied to a different type of substance is likely to be one reason why suggestions of 'other types of stuff' have been so fiercely resisted by atheists, yet it is not necessary for God to be an inevitable outcome. If there is such a thing as 'the stuff of Thought' then there are a range of possibilities about what it might bring in terms of additional capabilities.

A second underlying substance is likely to take many forms with differing levels of sophistication, in much the same way that Matter/Energy can range from fields and sub-atomic particles, to sophisticated planets with eco-systems.

In terms of achieving key aspects of our reality, the mechanisms we considered had to:

- Provide enough Matter/Energy to generate our Universe, (and possibly a succession of earlier Universes as part of an eternal cycle).
- Provide an additional source of Energy to power the accelerating expansion.
- Achieve the size of visible Universe, (93+ billion light years), within the timeframe of our Universe, (13.77 billion years).
- Achieve a very even spread of galaxies across the universe.
- Achieve the pattern of movement displayed by galaxies where their speed seems to be higher the more distant they are from us, (greater redshifts).
- Have a configuration of reality that supports sophisticated life.
- Provide uniformity – the same Laws of Physics applied to the entire Universe.
- Achieve all of this without resorting to luck – generally by providing an eternal cycle of experimentation; a single possible output; or a design.

It was surprising to me that the 'even spread of galaxies' and the 'pattern of their movement' are so very hard to explain in any other way than an explosion, which is why people had to think of mechanisms which could generate that scale of explosion. To date we have only conceived of 3 types of effect which might generate that pattern:

1) an 'explosive' push/release of energy from within the Universe.
2) an absolutely uniform force of attraction/suction surrounding the Universe.
3) distortions in the fabric of space.

An explosive push could represent various types of underlying mechanism from chemical/nuclear processes; to the outpouring of material from a hole between dimensions; or even a Framework for Existence if it was primed with new configuration settings. However at a more basic level we can also distinguish between a single huge explosion and a myriad of smaller ones going off at the same moment across a spread-out universe. The significance in this is that a very **broad** starting point **wouldn't** need to break the Laws of Physics in order to achieve the size of Universe we appear to have.

The 'splitting of nothingness'; a process of Induction; the transfer of material into our 3 dimensions from a parallel environment; or a change in Framework configurations, could all conceivably operate across very broad areas instead of from a small point in space. If such widespread effects occurred at the same moment across tens of billions of light years, they may require co-ordinated actions, and ever since the discovery that instant communications **are** possible across the universe[19] (even if we don't quite know how this is achieved), such co-ordination has become feasible.

If we move on to consider the idea of a 'suction' effect from a vacuum outside the Universe instead of an explosion from within, we should remember that science tells us the power of a vacuum is **not** normally generated by the void itself: it actually comes from the place where there is more stuff because that is where the greater 'pressure' will be generated. While this is certainly a practical truth here on Earth, and a good pointer to what may be true elsewhere, there is a force of attraction that operates on a cosmic scale - gravity.

The difference between this and true suction force is that gravity also comes from place where substance exists – not a void. We would therefore be suggesting that any existence beyond our Universe is not empty but filled with other material that is evenly spread around us, thereby exerting a gravitational effect on our Universe – pulling the galaxies outwards.

The idea that there is a true suction force out in space, in the void beyond our universe would imply that the Universe is truly infinite and does contain a force if nothing else.

The final possibility, (option 3 above), relates to distortions in the fabric of space. In the same way that Gravity (a force of attraction) has been likened to a heavy object sitting on a sheet of elastic with the resulting localised dent drawing smaller objects towards it as they passed close by, we could suggest that an entire Universe could distort the broader fabric of space into a slope down which the entire Universe is slipping – an additional force of expansion. In this analogy the Universe may be continually stretching the fabric of space until it breaks – a new form of Big Bang?

The alternate idea is to argue that gravity is explained by other means, (such as Steven Hawking's idea of negative energy), and that the normal distortion in the fabric of space caused by the presence of large objects is to cause 'a hill in the elastic' rather than a dent which would exist where there was a void. It is one way to explain both a suction force and why 'nature abhors a vacuum' and always tries to fill it.

I have a number of objections to this type of thinking, as explained in Chapter 8, which include the fact that in space there is no up or down and that the fabric of space is likely to surround an object – meaning that it will be stretched in all directions and not just create a hill. However as we don't know what such a fabric is capable of, it is premature to say that it cannot have specific properties, even if it does exist.

The other major discovery was that the expansion of the Universe is accelerating and that it began to do so 5 billion years ago, just before the formation of our solar system, (some 8.77 billion years after the Big Bang). To me, this is very significant because it forces three groups of issues into the debate;

a) in relation to the mainstream Bang-Crunch theory of origin, the acceleration would have ended the eternal cycle because gravity would no longer be able to collapse the universe into a singularity again... and the ending of an eternal cycle can only be done with a spontaneous or **ex**ternal factor.

In addition, the cycle itself could no longer be regarded as eternal so it couldn't fulfil its prime purpose of allowing nature to explore every permutation of existence to guarantee that our version of reality would emerge without luck. The only way to salvage this situation is if the cycle was guaranteed to **only** end after it had achieved our Universe as we will consider more fully below.

b) although new theories of origin are emerging which are immune to the acceleration effect because they use a different eternal cycle, (eg. the curtains of energy idea per the Epkyrotyic Model[38,41]), the sudden **starting** of the acceleration still has to be explained as it represents a fundamental change requiring a major cause to trigger it evenly across the whole of the Universe, (which in many scenarios requires a degree of co-ordination). Does it mark the tipping point in an inevitable process, or an **ex**ternal or spontaneous event?

c) there is a need for a source of energy to power the acceleration, beyond what was active before – and there only seem to be three generic possibilities for this, (the same mechanisms we had to explain the original Big Bang, labelled '1-3' above). However in terms of powering an additional acceleration 'from within', we can refine our scenario with a small number of sources that seem credible :-

- Some energy generated at the Big Bang was effectively archived for 8.77 billion years in a form that is still spread evenly across the universe, and that from 5 billion years ago it has been steadily converting back into dynamic energy. Perhaps there are scattered pools of Nothingness around the Universe, or possibly Dark Matter is being changed into Dark Energy?

- The additional energy must come from a new and enormous source outside our Universe, (such as other dimensions, a 2nd type of stuff... or God), which must also deliver that Dark Energy everywhere across the Universe at the same time, and without any apparent delay or major flow across space.

- The additional energy represents a new and sustained burst of pure creation across the whole of the Universe, not at a single specific point in space.

With all of the above factors in mind we stepped through each permutation of starting condition & mechanism to identify those process combinations which represented strong or reasonable possibilities for the origin of our Universe.

As part of this assessment it became clear that no suggested Big Bang mechanism had any proof of its existence, so this was not a point of distinction between them.

> Materialism had once been the only philosophy which could back almost all of its main theory of origin with firm evidence, but that is no longer the case. To this extent Materialism is a faith as much as Dualism and Idealism.

In terms of the results I achieved from my analysis of the different process combinations, it was interesting to see that the basic Materialist context, (without the influence of physical beings), had only **one** strong option to explain the origin of our Universe, and that relied on the swapping of eternal material with other hidden dimensions of existence. There were just 2 others which each had one serious concern about their viability which needed to be resolved, and these were also based on eternal existence. There seemed to be too many problems with ideas involving pure creation to make them likely possibilities.

Yet physical beings **are** within the Materialist concept, (because we are here), so it **is** conceivable that an advanced civilization in an earlier Universe might be able to influence the last Big Bang, as we saw earlier. However logic also suggests that any attempt by an advanced civilization to engineer a designer Universe is likely to result in their own destruction because it would fundamentally change the configuration of material that formed their bodies, so they would have no incentive to try such a thing. Although they might trigger a change by accident, such a scenario would put us back into the realm of luck, so it is much more likely that such a civilization would intervene to **prevent** a recurrence of the Big Bang in order to preserve themselves.

If that were the case then their intervention would be expected **after** the last Big Bang and indeed, starting an acceleration in Universe expansion would be a classic way to achieve this within the Bang-Crunch model - if that civilization had the means to do so.

Whether or not this has any bearing on real events, a principle we might draw from such ideas is that it might be acceptable to have a *finite* cyclical process as a way to avoid luck in generating our version of reality **if** that cyclical mechanism could *only* end once an advanced civilization was achieved. A 'mechanism' that could guarantee such timing might operate in the following way :-

> If living beings are the only perceived source of spontaneity, they may offer the only way to break an otherwise eternal cycle of universe creation, and their ability to end the cycle (in order to prevent their own destruction) would coincide with their achieving the necessary sophistication to do this.

The trouble is that physical beings have limited knowledge and are therefore prone to making mistakes, so strategically they only improve two options in our table, but they also make one worse, and they don't generate any new 'strong possibilities'.

It is only when we introduce the possibility of a 2nd underlying stuff, (primarily as a source of spontaneity/randomness), that the number of strong possibilities rises to 12.

Within these, the one strong Materialist possibility that I identified earlier remains viable, but the other 11 options have an advantage because they do **not** require other hidden dimensions to exist, although in fairness they all rely on other unproven factors. Whether that stuff is linked to Dark Energy, Thought, other characteristics, or even all of these things; as a source of additional capabilities that are discreet from the normal workings of Matter/Energy, they are potentially an invaluable concept.

The question is whether the characteristics we perceive are genuine or an illusion.

We seem to be drawn to the notion of pure creation because we instinctively like the idea of beginnings. Before I started this exercise I had the impression that if pure creation was possible it would answer all questions about the origin of existence, so it came as a surprise to find that various issues made it unlikely to be viable in a number of starting circumstances per our tables, (even if we said that a 2nd type of stuff could provide the capability to be spontaneous), because it was likely to:-

- be dependent on a specific conditions before it could occur, (such as a perfect void), and therefore some starting circumstances would be unsuitable, **and**
- be a one-off BB event and therefore dependent on luck to achieve our outcome, **or**
- produce different forms of reality if there were separate occurrences; going against the uniformity that we see across the cosmos.

Divine influence was needed to make pure creation a fully viable possibility through the application of design... and if you don't believe in God that would probably mean that you are also dubious about pure creation as well. However that doesn't mean that non-God versions of pure creation are completely invalid. They have to find ways to overcome some challenges.

One step in that direction has been to link the mechanisms of pure creation and 'Splitting Nothingness' to the idea of producing equal amounts of positive and negative matter as a way to 'balance the equation of creation'.

Although this might be viewed as a mathematical nicety, (because it **is** conceivable that pure creation may produce much more positive than negative matter), it can be argued that the nature of balanced opposites means that they would be more likely to emerge in only one form, (only one configuration of existence). If this were true, (and I'm not sure that logic supports this conclusion), it would imply that our version of reality would be the only one that was possible, so the 'luck factor' could be avoided.

Even if we argued that pure creation had led to an infinite number of universes, each with a unique configuration and that this would be guaranteed to achieve our version of reality somewhere in existence, it doesn't explain why our universe is the size it is; or why the accelerating expansion began 5 billion years ago; where the additional energy is coming from; or why this new stuff is in the same configuration of our existence if it does represent a 2nd prolonged burst of pure creation in our space.

There can be ways around some or even all of these challenges if we were to say that pure creation doesn't have to emerge from a tiny point in space but across a vast area; that it may be a prolonged process rather than a momentary flash; that it can be independent of all circumstances, (ie. be truly spontaneous); and without limits on the amount of material it can generate. Yet we also have to consider whether all of these additional requirements stretch the concept beyond credibility.

Finally, when stepping-through the different combinations of starting condition and mechanism, I used the concept of God in two ways: firstly as a way to illustrate the maximum that could be achieved from each of the natural mechanisms if the starting circumstances were arranged perfectly. Surprisingly, on this basis, only 2 circumstances, (4 combinations), became fully viable. These are the Bang-Crunch theory, and the splitting of a singularity made of 'Nothingness' into balanced opposites.

The second approach was to acknowledge that a Creator God could, of course, achieve anything either by building a new mechanism that was designed for the task, (but that would not represent a natural mechanism), or simply commanding it to occur.

Of course, you have to believe in the existence of God which would imply that God is made of something, which may be something like the intangible 'stuff of Thought' or potentially something quite distinct to reflect the unique properties of God.

Unusual Aspects of Our Current Reality – The outputs from Creation

Some aspects of our reality which stubbornly defy explanation by science are intriguing.

Firstly we don't know if Time is something distinct in its own right, (eg. a force of movement), or merely a symptom of something else, (eg. Matter/Energy; some other type of stuff; or a structure similar to another dimension). What we can say is that mathematics and some experimental evidence show that the speed of Time may not be constant - it can change in some circumstances generally associated with fast movement.

The distinct nature of Life, (ie. the underlying factor that changes a bunch of sterile chemicals into a living thinking body), has also remained elusive and this is covered in Book 2. However we **can** say that living beings do exist, and that Thought is very real. We use it every day, and Thought is the very thing by which we define ourselves.

It must therefore feature in Materialist thinking, but that philosophy has failed to provide a satisfactory explanation for the most interesting features of Thought – it just dismisses those characteristics as an illusion or says that they must surely be due to ordinary factors that have escaped our notice. It's not satisfactory even if you believe it.

Yet we have also failed to explain a core factor in our physical environment: the nature of 'fields of influence', (as opposed to the solidity we see in objects). Scientists have strong evidence that forces are transmitted by issuing minute particles from a source which act as messengers on how to behave to any objects that pass through the field. For various reasons described in Chapter 5 this implies that such influence is not necessarily exerted through the 'force of direct contact' or chemical reaction but potentially through an exchange of information – which in turn suggests that atoms have an internal mechanism that can make use of those instructions, (an internal Framework for Existence), or a degree of awareness – we're not entirely sure.

Unfortunately this theory has a couple of other serious challenges.

Firstly, science has only identified such particles for 3 out of the 4 known forces. Theory says that Gravity should be based on a 'Graviton' particle – but this has never been detected, which is strange because gravity is everywhere. (Even if recent claims about detecting gravitational waves are eventually shown to be correct, these effects are the symptoms being displayed by other things, not the Gravitons themselves).

Secondly, as a result of the Dual Slit experiments, some of those messenger particles (photons and electrons), may themselves turn out to be fields or indeed without any form at all – in which case an undefined field is being used to explain what a field is. In short – there would no explanation at all of what a field is.

I find it interesting that 2 of the main factors in physical existence which defy explanation, (Gravity and Life), are also the two factors which can be considered the 'assembling forces' within nature, rather than those which science says follow the 2nd Law of Thermodynamics in terms of increasing degradation, (entropy). While some scientists will deny the existence of any truly assembling force, (by saying that gravity is also a transient feature that will degrade with Matter/Energy), we have to recognise that a dynamic eternal existence needs to do things, which almost certainly requires both assembly and degradation in order to perpetuate that activity for all time.

Life is clearly linked with awareness, and in higher beings with Thought too. These are undeniable factors which must be accommodated within the Materialist perspective as much as any other. In other words, if Materialism is correct then Matte/Energy must be capable of generating awareness somehow... but in sub-atomic particles?

It is hard to ignore the fact that photons seem to be a common touch point in many circumstances which hint at awareness, because they are the 'messenger particles' for Electromagnetism as well as the prime particle used in the Double Slit experiments. If two photons collide they can also generate an electron... so it is possible that electrons may possess some of the communication abilities suggested of photons.

Wishful thinking? Perhaps not, (even if photons don't represent the full answer).

There are some facts that we cannot escape. Paired photon particles travelling away from each other at twice the speed of light, do retain an instant **communication** between each other which allows them to generate physical changes, (ie. allowing them to mirror each other's activities when they are many miles apart).

In addition, if you believe that photons or electrons are the only active elements in the Dual Slit and Quantum Eraser experiments, and that they remain as particles throughout, it is hard to avoid the suggestion that they must reach out, sense, and interpret their surroundings... a crude form of awareness, even if it isn't consciousness.

This is potentially taken even further in theories which suggest that photons & electrons are formless until measured or observed when they adopt characteristics relevant to their circumstances. As an example, in terms of them registering the 'which path information' at the moment of detection after passing through complex apparatus – that knowledge it is either something that is set during transit or something that must be derived at the end.

Of course there are other explanations of the results but they tend to require different assumptions about the circumstance & abilities of such particles. If you don't accept the possibility of crude awareness then you will probably favour one of these alternative theories which tend to suggest that particles either have remarkable properties which we haven't detected before, (eg. 'wave/particle duality'), or there are other types of stuff such as the 'hidden pool'.

However those who do favour the notion of awareness can point to the remarkable findings from the Double Slit, Gisin and Bell Tests experiments, (Ch. 6 & 7), which appear to support the QM view of the world, (ie. that particles take form when an outcome is required).

Strategically, in this context, I feel it is significant that **Thought seems to be the only thing which can cause Matter/Energy to deviate from its inevitable chemical path.** We see the examples in our own lives all the time. Cars and buildings would never arise without the influence of Thought.

In this respect it is also interesting that the Laws of Physics are completely **unable** to predict human activity, which adds a degree of weight to suggestions that Thought may be a source of spontaneity or randomness in the world – and being the opposite of cause & effect such characteristics *couldn't* come from Matter/Energy.

If you find characteristics that are not within the capabilities of Matter/Energy you have to position them elsewhere, and it is interesting that spontaneity/randomness; awareness; and sophisticated Thought, can be seen as increasingly sophisticated manifestations that might emerge from another type of stuff.

Of course, that doesn't make it true.

However when we consider how our Minds might work with our bodies, a valid implication from the suggestion that Thought is based in another type of underlying substance must be that there is a 'touch point' with Matter/Energy if physical material is to be influenced. Any action within physical matter requires an exchange of energy, and any influence that Thought could exert would be no different.

We do know that the brain uses electrical signals but it is unclear how those originate. Do neurons purely release chemical energy or is there a hidden exchange as well? Book 2 will explore these implications in more detail.

However the idea that intangible stuff may contain energy and may also provide a source of spontaneous creation, (possibly illustrated by ideas suddenly popping into our heads), has allowed some people to speculate that the pure creation of ideas could also be the pure creation of energy... and with enough energy enormous effects might be achieved – like a Big Bang.

It is hard to ignore the associations between this idea and religious concepts, which is why it is highly disputed, but it does illustrate the point that pure creation may only apply to intangible things. This possibility brings pure creation closer to our own circumstances but it has a long way to go before that idea can be regarded as a serious possibility. There are too many unknowns.

Conclusions

I was in a dilemma about this final section because although I would like to draw some conclusions, they would be mine when they should really be yours. As I promised you at the start of this book, I am **not** here to promote any view but to present you with the range of thinking. So my aim here will be to provide some focus on where the debate may go next.

One of the things that I can safely say is that as things stand, **there is no single conclusion**, but there are **a narrow range of possibilities** which seem to fit the facts that we have discovered so far. My personal assessment of that range, (across all beliefs), was shown in the tables of Chapter 12.1 and in Appendix A.

Where we choose to position ourselves in that range depends on how we each frame the issues surrounding existence. What are the rocks on which we base our personal perception of reality? Where do we draw the line on 'flexibility of concept' when facts contradict our expectations?

To test your initial thoughts on that positioning, consider the 4 stand out findings from science that we have considered and which still need to be explained:-

a) the visible appearance of the Universe – in terms of its age, size, the even distribution of galaxies, and the distinct pattern of their movement.

b) the Double Slit and Quantum Eraser experiments which challenge the nature of particles and even introduce suggestions of time travel.

c) the results from the Gisin & Bell Test experiments which suggest that paired particles have a sense of mutual awareness and influence each other instantly.

d) the accelerating expansion of the Universe – in terms of how it started and the source of its power.

We also need to form an opinion about the fundamental nature of existence:-

1. Are you a robot, or a person with Free Will? Are your ideas/decisions/actions inevitable, or something to be applauded /criticised, (creative & spontaneous)?

2. Is physical substance real or an illusion? Is existence merely a trick of the Mind?

3. Does space have to exist before it can be occupied, and if so, does this mean that there has to be a fabric or framework to existence to impose its rules; establish the number of dimensions that can be occupied; and enforce uniformity?

4. Is Time something in its own right or merely a symptom of something else – the rate of change in other things that exist, (Matter/Energy or other types of stuff)?

5. Which bits of existence (if any) are eternal and which had a true beginning?

6. Can the Laws of Physics be broken, and if so, under what generic circumstances?

In their different ways these questions will guide you on some more detailed points concerning the nature of any existence prior to the Big Bang and whether you feel that there is either evidence or a logical need for capabilities that go beyond Matter/Energy:

7. How many types of substance underpin reality; what are their distinguishing characteristics; did they have separate origins or did they always co-exist?

8. Which type of Thought do you believe in: 'Physical'; 'Intangible'; 'Pluralistic'?

9. Are we aware of *any* events across the entire history of existence which mark a fundamental start or end point, (ie. **that were not inevitable)**?

10. Just before the last Big Bang was the physical material in the Universe 'zero'; or concentrated into a tiny space; or already widely spread?

11. If enormous structures/mechanisms exist within the Universe, to what extent might an advanced civilization of **Physical** Beings, (entirely based in Matter/Energy), influence the configuration of reality?

12. How can your emerging view of existence resolve the six impossibilities I listed in Chapter 3? Can we avoid them all, or must some, in fact, be possible?

In re-visiting your beliefs based on these 12 questions you will hopefully be able to re-set or re-confirm your stand point based on the latest facts and thinking that we have available... but there is still a choice. Beyond what is known, faith may lead us to hope for discoveries that can plug gaps in our personal beliefs, yet those could lie in very different directions. We should at least recognise what **is** known and has to be explained.

One of the most unusual aspects of this debate is that **all** explanations of existence say we are living in an illusion, and one of the personal decisions that we face is to decide how far the illusion might go.

In our daily lives we are forced out of practicality to operate on the assumption that physical matter truly exists, has substance, and operates according to rigid rules.

It's a very practical assumption yet the most basic truth of all is that our perceptions originate in the Mind.....   I think therefore I am.

When the Mind is the only thing we can guarantee to exist, it seems equally strange that the dominant philosophy of the modern age argues that Thought is the ultimate illusion; that we are all based in Matter/Energy; and also that the characteristics which we use to distinguish Thought do not really exist. It may be correct, but as things stand it is a faith that cannot prove its case. The stuff of Thought remains a possibility.

We should remember that Thought is real and must be accommodated within every philosophy of existence. We use it every waking moment, and that gives each of us an enormous amount of evidence which should be explained rather than simply dismissed out of hand. There are 3 main ways to do this as we saw in question 8.

Sadly, there has been a tendency, particularly for the 'monist' perspectives, (believing in just one type of underlying stuff), to dismiss large swathes of our experience as illusion rather than fact. We should be very wary when people do so.

Belief in a single type of stuff is a faith/doctrine. It is not fact... at least not yet.

Whether we look at scientific findings or our general experience of life & Thought, it is important to separate fact from interpretation, and then explain all facts. This is because interpretations can change, while facts do not - ever.

Unless somebody is lying, the records of an experiment will be a fact even if they are later shown to include error, (for instance, if the equipment used had been incorrectly calibrated leading to false readings). It's the *interpretation* of facts which will judge whether an error was made, or that the readings imply a particular type of behaviour.

Yet if different scientists repeat experiments and achieve the same results there is little chance that the raw results are in error. We have to focus on the new characteristics and determine whether they are a previously unknown part of Matter/Energy, or they represent something quite separate.

Equally, if our shared experience of how the natural world behaves is confirmed by almost everyone, there needs to be a good reason for suggesting that it doesn't work that way. If society believes that we have Free Will and that our choices are not inevitable, do we change society to fit the implications of strict Causality, (single cause leads to single effect), or do we look for other things to explain the random or spontaneous factors that would make us responsible for our actions?

To date we have allowed society and science to pursue different philosophies partly as a way of finding the boundaries for these arguments, but we are approaching a time where those philosophies have to be united in some form. Materialism is increasingly having to consider **ex**ternal factors as the only way to explain seemingly random or spontaneous events, (moments of real change).

Science has traditionally aimed its efforts at finding answers within the only stuff that we can detect, Matter/Energy, and that's no bad thing, however there comes a tipping point where old beliefs have to be challenged once the evidence for new capabilities in existence becomes powerful enough. In the case of spontaneity and randomness the evidence comes from quite a few sources such as

Thought, the actions of **sub-atomic particles,** the **logic of origin,** and explanations of **fundamental change.**

That tipping point has been reached in a few areas of science that are often associated with cosmology, the nature of existence, and issues of origin, (from cosmic inflation to explain the size of the Universe, and a power source for the acceleration, to hidden dimensions, Frameworks, and even a multiverse). The related organizations (eg. NASA and institutes of Astro-Physics), **are** now openly suggesting that there is a logical need for Dark Matter, and Dark Energy, if nothing else.

I also found it interesting that once Dark Matter was seriously proposed, new ways were found to provide greater evidence for this other stuff. That evidence points to symptoms rather than directly to the substance itself, but it is still credible. Some proposals are also emerging which may ultimately demonstrate the reality of Dark Energy too, although this stuff currently remains theoretical. What could we achieve if we actively went looking for other things indicated by the evidence?

Yet there seems to be much greater reluctance to pursue such suggestions in other fields of research – such as in relation to explaining the Dual Slit experiments. The deBroglie-Bohm theory argues for a hidden guiding wave that had no energy, but could it be that it is just a type of energy that we can't yet detect – equivalent to the hidden pool I described which *might* be linked to Dark Energy?

The reluctance within Quantum Mechanics to pursue this line of thinking seems to be because Dark Matter and Dark Energy can be likened to normal Matter/Energy – the rationale being that while distance may prevent us from detecting this pseudo-normal stuff in space, we would expect sensitive equipment here to detect such material, but it doesn't. However, to make a point, we can't detect gravity either.

We therefore face a choice: do Dark Matter and Dark Energy represent something new which we can't detect, or do the findings from the Dual Slit experiments have to be explained without them?

To help focus our search for solutions I also presented some perceptions about our reality which have not received any serious challenges, but they need to be considered further before they can be considered robust.

One that may have particular significance is that: *spreading waves are only ever produced when objects pass through a pool of other stuff,* (like a train through air, or a ship crossing a lake or sea). If correct, then whenever we see a spreading wave effect it would be natural to look for the pool rather than invent new properties for the objects passing through it.

As another example, I haven't found anything to disprove the suggestion that

Thought is the only thing that can cause Matter/Energy to deviate from its inevitable chemical path.

If you **can** think of another example it may help to unlock the mystery of Thought, however if it is true then Thought must, at the very least, be on a separate path to Matter/Energy, and quite possibly something distinct from it.

While it can be argued strategically that Physical Thought, (based entirely within Matter/Energy), might somehow be on a separate causal track to the physical environment and therefore able to act as an **ex**ternal influence, we would also have to see if this explained all of the scenarios which indicate randomness/spontaneity in decision-making.

If we find a single example where Thought does not operate on the basis of strict causality it must break the principle and force us to consider other factors.

The properties of spontaneity and randomness have long been suspected but are robustly challenged by those who say that these characteristics must be false because they defy rational explanation, as well as the fact that we can't detect any 'stuff of Thought' – only the electrical signals in the brain. Yet that is tenuous logic.

To say that **those characteristics cannot be rationalised on the basis of causality** is inevitable because they are the opposites of cause & effect, so that seems a rather unfair basis for dismissing these suggestions out of hand. Equally, Determinism cannot explain a large number of actions attributed to spontaneity or randomness, so it cannot claim to have resolved the issues.

Electricity doesn't think, and science cannot explain why a neuron would fire just at the right moment without a physical cause. (Neurons do not fire like clockwork to give us a new idea 'every second on the second' and brain scans show that when we have an idea it can spark activity in separate parts of the brain that do not seem connected - there isn't a flow from one part of the brain to the next in a physical sequence).

In overview, *the absence of evidence is not proof of absence*, so at what point do we say that the symptoms are credible enough to overwhelm belief in a doctrine?

The potential evidence for spontaneous/random behaviour is also no longer confined to discussions about Thought. It is emerging with the realisation that Materialist explanations which rely on 'chance' or 'mistakes' are not satisfactory, and in the various findings of Quantum Mechanics which throw an increasing spotlight on the behaviour of sub-atomic particles. From the new forms of reality which supposedly emerge from repeated explosions of **the same singularity**, to supposed **errors** being made by the sophisticated DNA repair mechanisms in each cell; proper causal explanations are needed

Cause & Effect doesn't accommodate Chance or Mistakes, it only allows for one inevitable outcome.

Against this background I hope that you can see that the ties back to philosophy are quite strong. I say again that Determinism and Materialism are faiths as much as Dualism and Idealism. They all have to prove their case, and to say that one is the truth when it is unproven, and the other options haven't been properly considered, suggests to me that these claims are premature. But of course, these are matters of personal choice and you may well take a different view.

At first glance, from the above narrative, you may say that the philosophy of Idealism has little to back it up, but closer scrutiny may change your opinion.

The suggestion by scientists that things such as photons and electrons **have no shape or form** until they are observed/measured/detected is very similar to the concepts

211

within Idealism. They are often driven by assumptions that if there is no external physical influence, the adoption of one state or another *which is also relevant to the circumstances* must come from the photons/electrons themselves. If they respond to the environment that they encounter they would either have to be given controlling information about their circumstances, (prompting them to give an inevitable response), or they must reach out to determine and interpret those circumstances... say a crude level of awareness about the certainty of the 'which path' information.

I am sure that many people will scoff at any suggestion of sub-atomic particles having awareness, even in the simplest/crudest degree, however, that description doesn't seem unreasonable in view of the non-physical link that paired particles clearly seem to demonstrate. It is even suggested that this extends to the **retrospective** alignment of dead particles at D0 in the Quantum Eraser experiments.

Of course, these are interpretations rather than the raw facts, but as the Bell Test experiments were designed to disprove the challenges to Determinism/Causality, the unexpected opposite result must give added credibility to other beliefs - including those that involve randomness & spontaneity.

We don't know what has yet to be uncovered, and it may be that new factors can radically change our perception of the experiments. However, as things stand, if a theory fits all the evidence then it must be considered a valid option.

So it becomes a matter of faith – whether you feel that there may be other hidden factors underpinning existence. For instance, if you believe that sub-atomic particles can display crude awareness, how might that have helped to shape reality in the early Universe? If you believe in other types of stuff, how might that have evolved?

Other types of substance might simply provide capabilities for spontaneity or randomness. They don't have to imply Intangible Beings or God, but equally it **could** be one way to explain such beings if they were shown to exist. However at present, that isn't the case. Their involvement is theoretical and related to scenarios of origin.

As I suspect that most people believe in the reality of physical substance, most would incline to the view that the only examples of life we encounter are either Physical (based in Matter/Energy alone), or Pluralistic, (based on a close dependency between Matter/Energy and the stuff of Thought).

With this lead-in I would then like us to consider a keystone in scientific thinking - the 2nd Law of Thermodynamics which essentially argues that all physical matter will continue to degrade in a process which will ultimately lead to existence becoming one uniform 'background' state. This has been used a way to explain why Time will only move forwards and not backwards... and yet it ignores two important factors

- in order to degrade, something must have been created/ordered/structured in the first place. The two obvious candidates for doing this are Gravity and Life and these are both **un**explained by science.
- most Materialist explanations of origin are based on eternal existence, and if there were no continuous forces of construction then we might expect that existence would already be bland, uniform, and fully degraded.

The achievements of science in accurately predicting the actions of physical material within our Universe will remain valid even if the context broadens through later findings, so it may be that the 2nd Law only applies once things have been constructed. Yet without these limits it's hard to reconcile the 2nd Law with theories of origin.

The appeal of beginnings, and the apparent beginning marked by the Big Bang, have led scientists and mathematicians to find a way to put pure creation into a balanced equation. The concept is that either through pure creation or the splitting of 'nothingness' balanced quantities of positive and negative Matter/Energy will emerge. Yet the reality that we face day to day is overwhelmingly 'positive'. There are only 3 generic ways to resolve this imbalance if the theory has any bearing with reality:

- negative Matter/Energy is **accumulated in other parts of existence**, (either other parts of the Universe, or in places beyond our Universe).

- negative Matter/Energy is explained through **other features**, (eg. gravity is negative energy).

- nature is formed mainly of positive Matter/Energy and **there is no balance**, which either means that pure creation doesn't 'balance the equation', or existence has simply been that way for all eternity.

Interestingly, the main potential evidence we have for pure creation are the new ideas we generate. Almost all of the things we normally point to as having a beginning can be traced back to human thoughts and the designs/initiatives we generate. Could this mean that pure creation is only achieved through the intangible and spontaneous stuff of Thought? If so, the outputs from creation are also likely to be intangible and in terms of prime origin this intangible stuff would then have to be converted to physical energy and then physical matter by some means. It's a nice idea but there is no real basis for making the call one way or the other, except through faith in a philosophy, Materialistic or otherwise.

The proven reality that paired particles can instantly mirror each other's activities despite moving in opposite directions at the speed of light, is a stunning finding from science. It is direct evidence of something that directly contradicts traditional scientific theory about the limits of the speed of light, and the choice is either to junk that part of scientific theory, or find **additional** factors in nature to explain it.

While ideas about hidden structures remain entirely theoretical, (whether we are talking about the Fabric of Space, Frameworks for Existence, hidden 'Curtains of Energy', etc), the one concept which may be supported by those instant communications is the possibility of hidden dimensions of existence.

However, if you feel that such structures are the only way to explain some of the symptoms we have been recording, then you also have to consider how they arose. These structural elements would be fundamental to existence and would have to appear very quickly after a Big Bang if they were to have any effect.

Unless we venture into the realm of God I find it impossible to think that things such as a Framework for Existence could be retro-fitted by design. These structures are not simple things either. What could be organised enough to fashion them in the midst of explosive chaos in the early Universe?

We might also ask why unthinking chemicals would create them when they wouldn't care whether anything existed? Yet without them, new explanations may be needed to fill a gap in the factors that we know about.

Our awareness of existence is growing from the middle outwards, and we only have a small number of 'rocks' on which we base the foundations of our understanding. Any new bit of robust evidence that adds to our perceptions must be welcome. Equally, those scientific principles which have accurately predicted the future for some time should be retained.

Given how few of these rocks there are, I find it surprising at how easily some people are prepared to dismiss them – almost on a whim.

As religions have discovered, it is very dangerous to make pronouncements based on doctrine rather than facts, and the same danger looms for scientists today if we close our minds to certain possibilities that are strongly indicated by the evidence. It doesn't make them true but it shouldn't rule them out either without proper reasons.

It may be that 'strangeness' puts us off some suggestions, but the vast majority of ideas about origin are strange, and it is often the rule that the simplest explanations are the best and most accurate even if they contradict some established wisdom.

I hope that with greater awareness of the possibilities and an open mind we can break down barriers, make better decisions, and enjoy the developments/findings to come.

Appendix   A

Technical Review : Viability of Arguments for Origin

A1   Criteria for a Big Bang Process

In Chapter 8 we identified a number of conceptual factors that may have shaped a Big Bang type event. In overview these are shown in the diagram below, but in narrative form those factors are :

- The **starting conditions**, (ie.the state of any existence immediately prior to the Big Bang which also provide the raw materials for such activity).

- Various **mechanisms** by which the Big Bang could have operated, (plus the associated triggers that could activate each).

- Various types of temporary or permanent **enablers**, representing configuration settings, frameworks, or process to achieve our particular outcome.

In addition, we need to ensure that those things which emerge from the process have the exact characteristics of our observed Universe, many of which will have a backwards push onto any explanation of the nature of the Big Bang itself.

The Conceptual Big Bang Process

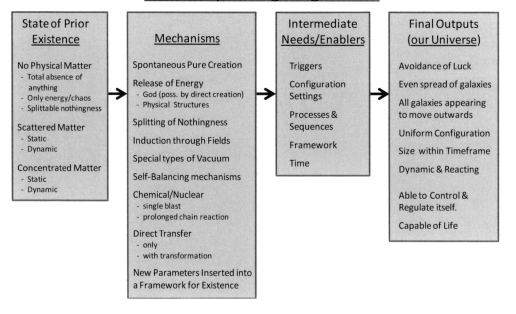

State of Prior Existence	Mechanisms	Intermediate Needs/Enablers	Final Outputs (our Universe)
No Physical Matter - Total absence of anything - Only energy/chaos - Splittable nothingness Scattered Matter - Static - Dynamic Concentrated Matter - Static - Dynamic	Spontaneous Pure Creation Release of Energy - God (poss. by direct creation) - Physical Structures Splitting of Nothingness Induction through Fields Special types of Vacuum Self-Balancing mechanisms Chemical/Nuclear - single blast - prolonged chain reaction Direct Transfer - only - with transformation New Parameters Inserted into a Framework for Existence	Triggers Configuration Settings Processes & Sequences Framework Time	Avoidance of Luck Even spread of galaxies All galaxies appearing to move outwards Uniform Configuration Size within Timeframe Dynamic & Reacting Able to Control & Regulate itself. Capable of Life

The analysis that we're about to conduct will consider each combination of circumstance and mechanism to test its viability and likelihood – initially on a purely materialist basis (without Thought) then with different levels of added influence from thinking beings. In summary this means 4 exercises :-

- Without any thinking influence.
- With the influence of thinking *physical beings* like us, who can only use Physical Thought that is based entirely in Matter/Energy.
- With the possible influence of a 2nd type of underlying stuff related to Thought, (ie. separate to Matter/Energy).
- With divine levels of Thought, guidance, and power.

None of the mechanisms are without their flaws. If nothing else, none of them have any scientific proof that they actually exist, however some are elaborations on mechanisms which do exist on a smaller scale within our environment, (such as induction), while others require a stronger leap of faith, (such as the presence of 'Curtains of Energy' or God).

For this reason we have to assess them first on the basis that they **are** possible in the right circumstances. By this I mean that certain mechanisms require pre-conditions. As a simple example you can't have a chemical or nuclear reaction in the absence of Matter/Energy or if everything is entirely without movement. Nor can you have a rip or tear in the fabric of space unless space has a structure of some sort.

So the starting conditions have to be compatible with the mechanisms being proposed. Those which aren't compatible must be excluded as impossible – ie. a combination that is not applicable.

The obvious way to ensure that we cover every permutation is to work our way around a table that lists them all within a grid or matrix. Once we have conducted the exercise on a purely materialist basis then we can set up the same table but imagine the level of success for each combination with the added influences of Thought.

There will therefore be 4 sets of tables which we can then compare at the end.

In the coming pages I set out one set of results with an accompanying set of notes to explain the logic which led to each assessment. It is intended to act as a guide in the hope that you will conduct your own analysis and reach your own conclusions.

Different perspectives will almost certainly lead to different conclusions on some process combinations, but I hope that in overview the level of difference won't be too great. However that is for you to decide.

An example of the main section of the table is shown below. On the left you will find a list of possible mechanisms as identified from the narratives in this book. At the top you will find the headings which summarise each type of physical environment that may have existed immediately prior to the last Big Bang.

This version of the table shows the materialist perspective without any Thought, but in later versions where Thought is layered on top, we have to imagine which columns could include each type of Thought. For instance, physical beings can only be present in a dynamic environment with scattered matter, rather than a singularity.

Starting Circumstances – Matter/Energy Alone

Ref	Mechanism	a. Total Absence — Nothing physical in our 3 dimensions	b. Absence of all Matter — permitting unstructured energy & forces	c. Scattered Matter — Static (eg.without Time)	d. Scattered Matter — Dynamic	e. Concentrated Matter — Static (eg.without Time)	f. Concentrated Matter — Dynamic
1	Spontaneous Pure Creation	L	HU	HU	HU	L	L
2	Creation by God						
3	Energy Release - Structures			X			
4	Splitting Nothingness		L/S	X	L	X	HU
5	Induction		L/S	X	L	X	HU
6	Vacuums/Voids in space		X	X	HU	X	X
7	Self-Correcting Structures		X	X	L	X	HU
8	Chem./Nuclear - 1 explosion		X	X	HU	X	L/S
9	Chem./Nuclear Chain Reaction		X	X	L	N/a- duplication	
10	Direct Transfer - Channel/Tube		X	X	HU	X	HU
11	Direct Transfer - Dimensions	L	L/S	X		X	✔
12	Frameworks & Parameters		X	X	L	X	L/S

Each cell in the table represents a combination of a particular mechanism with an opening circumstance. The colour and letter which I then apply indicates my assessment of the likelihood of that combination successfully producing our exact Universe, based on the logic in the notes. The colours and letters which I deploy are as follows:-

	✔	A strong possibility that may even be 'Likely'
	P	A reasonable possibility with 1 main caveat/additional condition or circumstance to apply.
	MF L/S	These circumstances are unlikely to be successful because there is more than one problematic factor with them which may be because they require luck or spontaneity to generate our Universe.
	U X	Highly Unlikely to be viable even if the mechanism and circumstances do exist or not required
	N/a	The combination of Mechanism & Circumstance is not applicable – ie cannot occur.

In a few rare circumstances certain combinations will represent a duplication of the proposed process and I have excluded it as each option should only be considered once.

From the notes you will hopefully also see that in order to achieve a consistent approach which minimises personal bias, I have labelled processes as a '*strong possibility*' where there are no significant issues beyond the existence of the relevant mechanism; a '*modest possibility*' if there is one additional caveat; and an '*unlikely possibility*' if there 2 or more additional caveats.

The 12 mechanisms which I use are a summary of the possibilities considered in Chapters 10 & 11, which can be described as:-

1. **Spontaneous Pure Creation** of Matter or Thought including the very 1st idea.

2. Release of **Energy from God** or **direct creation by Him**.

3. Release of **Energy from Physical Structures**, (eg. energy membranes; circle of universes).

4. The **Splitting of Nothingness** into Matter & Antimatter requiring high energy + a trigger.

5. **Induction** generated by the movement and interaction of 'fields'

6. A movement of pre-existing material to **fill a 'vacuum'**.

7. A movement of material due to **self-correcting mechanisms** within a structure of existence, which may also cross hidden dimensions of existence.

8. A **Single Chemical/Nuclear Blast** that spreads material in all directions.

9. A **Chemical/Nuclear Chain Reaction** where the outputs from one reaction overwhelm other material further away triggering it to react in the same way.

10. **Direct Physical Transfer** of material or energy **within our 3 dimensions** from one place to another via a hole, channel, tube, or path.

11. **Direct Physical Transfer** of material or energy **across hidden dimensions** of existence which also transforms matter from one configuration to another .

12. The **Insertion of New Parameters** for existence via a 'Framework for Existence'.

Aside from Pure Creation (rows 1 & 2), all of these mechanisms require a prior existence which is also assumed to be *Eternal* in order to effect a Transformation of existence. In this way we cover the 2 basic start point options.

The Starting Conditions should be relatively clear however column 'b' could accommodate a variety of circumstances from drifting raw energy, (the equivalent of 'chaos' in ancient texts, or dark energy in modern science), to nothingness as the 'zero form of something' which can be split.

If eternal physical matter exists its state can be flexed in two ways:

- In terms of movement, it can be imaged to be completely static or dynamic
- In terms of location, it can be regarded as concentrated in one place (a singularity) or scattered across a vast volume of space – a universe without singularities.

These permutations give us 4 options for the original state of Matter/Energy.

I have chosen to show Time as an 'Enabler' in case it does represent a distinct base element of reality, however in terms of its effects on the mechanisms of the Big Bang, the only variable is whether or not Time is present. It can be assumed not to be present in an environment where there is absolutely no movement or sequence, (ie. static conditions), but we know that it has to be present by the time our universe emerges.

For these reasons and in order to save space I haven't provided an additional column for it, and will also consider the other enablers within the circumstances of each cell in the table.

A1.1   Necessary Conditions for the Enablers

The 'Enablers' which I listed need to be present at specific moments relevant to each suggested Big Bang process; *which may be* **before,   during,** *or* **after** *the Big Bang moment*, in order for the whole process to work.

As a refinement on this thinking, and in relation to certain circumstances, some enablers may need to be generated before the Big Bang but only applied afterwards.

We need to bear those sequence in mind, regardless of whether we're considering the circumstances of pure creation or of transformation.

Triggers can be physical or circumstantial. As a simple example we can think of a trigger mechanism for a bomb that is either based on a clock or the pushing of a pin as the bomb hits its target (*physical*), or waiting for the right *circumstances* to arise, eg. it fires when it reaches a certain air pressure – ie. a certain height. A more subtle type of *physical* trigger could be when a living being decides to act for whatever reason,

All '*circumstantial*' triggers fire because the necessary conditions happen to arise through the movement of existing things, (resulting in collisions, vacuums, pressures, or even the simple volume of material). For instance huge gas clouds in space can suddenly ignite and become stars simply because enough material has been gathered and compressed together in one place through the natural action of gravity, or in the case of uranium if you simply produce a big enough lump it will become highly radioactive as it starts to react with itself .

While physical triggers need to be put in place, circumstantial triggers do not: they just happen. Yet circumstantial triggers do require movement, (ie. they can't work in a static environment).

The following items are a short explanation of the logic which can help us to determine a sequence and requirements for each situation, (combination of circumstances and mechanisms):-

'i'   by definition triggers have to be available at the moment that an event begins. All *physical* triggers will need to exist (and therefore be constructed) before the event, while *circumstantial* triggers are likely to come together at the moment itself. Both require something to exist before the Big Bang.

'ii'   If existence has to be configured, and stable/compatible settings have to emerge in some way, then without Thought/design we have to consider how our particular configuration of existence came about.

In the context of pure creation this could only take place after the new stuff has been generated and some of its properties begin to emerge, however in the case of eternal existence then there may be some legacy from a prior Universe if there is a cycle of change.

Depending on the mechanism by which these settings could arise we can take a view as to whether thinking beings who are not deities could exploit it to generate their custom version of the Universe.

'iii' As natural processes must reflect the characteristics of the things on which they are based, (ie. forces + matter), naturally occurring processes and sequences can **only** emerge, or be applied, **after** the other elements of existence are in place.

If those base elements had been fundamentally changed by the Big Bang event, then earlier processes would presumably no longer operate, (ie. they could no longer exist), so new ones would have to emerge that reflected the new properties of Matter.

In other words, the physical processes of our current Universe (ie. the Laws of Physics) could only emerge **after** the moment that the Big Bang was triggered and after a new configuration was applied.

The only caveat to this is that if it took a while for the Big Bang transformation to be applied everywhere in the Universe, some old processes and structures from a prior existence may have co-existed with the new, for a while – possibly long enough to interact briefly with the new and thereby influence it.

'iv' In relation to frameworks :-

A 'Framework for Existence', (if one exists), is likely to have pre-existed the Big Bang where physical matter existed without beginning, and also where pure creation emerged alongside pre-existing material. This is for 2 reasons. Firstly, the natural development of a Framework is likely to be because nature needs one, in which case the prior existence is likely to have needed it too. Secondly, it may be necessary for a Framework to exist in order to achieve uniformity across the Universe after a Big Bang event.

On the other hand, in the circumstances of pure creation, we have to consider whether there would be enough time for natural forces to develop a framework in the time since the Big Bang event. While it is developing we also have to ask whether it could influence anything.

Overall, this logic leads us to the sequence presented in the following table which also contains some finer-tuning to reflect whether things had to exist before, during or after the Big Bang :

Factor (presented in order of sequence)	Pre-BB At the Moment or After BB	Comment
Absolute Time Eternal Existence Spontaneous Creation	Pre At the moment	If it exists as something distinct, Time would be necessary for movement and sequence including the triggering of Big Bang. Time either began with the emergence of dynamic Matter / Thought, or with the 1st spontaneous movement.
Framework for Existence (if it exists) Eternal Existence Spontaneous Creation	 Pre and/or After -	A framework might exist without beginning but it only has purpose after Time begins as it is a structure to control dynamic matter/energy. Probably pre-existed BB if we believe that everything falls within its influence, but some scenarios do require a new Framework after the Big Bang event. A framework is probably unnecessary if a single process of creation generated a uniform output, and it is possibly unachievable as a direct output from a random process.
One or more Triggers Eternal Existence Spontaneous Creation	 Pre or At -	Must have preceded the Big Bang or have arisen at the exact moment that it fired. Spontaneity means no cause, so cannot have a trigger.
Our Configuration Settings Eternal Existence Spontaneous Creation	Designed Pre. Always applied at the moment of Big Bang At the moment	Whether they are designed or they emerged from inevitable natural factors, settings that were different to a prior existence could only be applied at the moment of the Big Bang. Design would mean that plans occurred beforehand The outputs from creation would be the result of luck and could only emerge at the moment of creation.
Physical Processes & **Repeatable sequences**	After	They depend on settings that emerge from Big Bang, so any development or implementation must occur after the event. They must be an inevitable consequence of the settings unless subject to later tinkering by God.

From the table above we can determine a broad sequence in which 'Enablers' might be applied to the various combinations of mechanisms and starting circumstances. However, while this is a relevant set of background information my attempt to use these factors to reduce the number of valid permutations has not borne much fruit.

The main reason for this is because not enough is known about the mechanisms being deployed. By way of example, we have no idea whether pure creation could directly produce a 'Framework for Existence' or whether any physical matter that was generated would already have various configuration settings.

With so little information we can only speculate about the likely need for certain things, and I will share some ideas with you which may have relevance in future.

Frameworks are unlikely to be required where :
- a single process will be applied in the same way across the entire environment, eg. chain reactions where the same outputs perpetuate the same cycle of activity.
- matter/energy that is released from a structured environment is likely to conform to the parameters which exists in that feeder environment.

Frameworks *may* be necessary where
- matter/energy is released in an unstructured way such as Induction, or where the format of material could be altered, or the splitting of nothingness could arise in different ways.
- a single process of pure creation cannot be guaranteed to conform to the same standards (ie. a variety of configurations might emerge from a single process), or where multiple creation events occur.
- God may or may not feel the need to create one.

Distinct **physical** *Triggers* are not required :-
- in the case of spontaneous physical activity,
- when activity is sparked by particular conditions/circumstances or induction,
- where activity is generated by God or spontaneous thoughts,
- where new configuration parameters emerge in a singularity, or
- where self-balancing applies.

Distinct physical Triggers *may* be required where activity has to be initiated outside the normal sequence of activities.

A2   Detailed Analysis

It's important to realise that there is no definitive answer to this type of analysis, so you are likely to achieve a different assessment to me. This may have relevance for certain specific options – for instance I may deem something to be a modest possibility because of certain issues, but you might consider it a strong possibility because you place a different level of importance on some aspect of the scenario.

I'm hoping that your analysis will only be subtly different to mine – reflecting slight differences in grading – but who knows!

However I feel we will gain as much from what the overall picture tells us as we can from specific process combinations. In particular it will be interesting to see how we feel that the chances of success will be improved by the introduction of Thought.

We may also get some interesting information from the pattern of viable combinations across each grid, in which case we may be able to get a sense of which factors may be more important to the debate than others.

I am concerned with prime sources of change, so if pure creation leads to a chain of other activity, (such as nuclear reactions with things that already existed), I will still consider the main mechanism to be the one which got things going.

I have also made assumptions about the starting circumstances. For instance I have supposed that a singularity is likely to lead to a much larger explosion than would be possible with scattered material.

In other respects, some of the mechanisms can only work *within* the Laws of Physics while others have the opportunity to change the balance within those rules, to create a different permutation of the same underlying reality. We have to remember the limits of this positioning.

These thoughts represent aspects of the 'bottom up perspective', however there are also challenges in ensuring that the end results of any process achieve our Universe – the 'top down' scenario. In this respect we are drawn back to some key observations about our Universe :

- Galaxies are spread evenly across the Universe, ie. there are no particular concentrations of matter.

- There is a universal pattern of movement across the Universe in that all galaxies seem to be moving outwards from the centre, and the more distant the galaxy the faster it seems to be travelling.

- The Universe is far wider than could be achieved by material travelling at the speed of light.

- All observed matter in the Universe seems to adhere to the same rules of existence as we experience on Earth, so we assume that it is all made of the same 'stuff'.

- If there was previously an eternal 'Bang-Crunch cycle' (per Ch. 4), the accelerating expansion of our Universe shows that this no longer applies to our Universe, and the change must have been the result of an external or spontaneous factor.

The other main issue surrounding the process of origin is to avoid luck in generating our outcome. If a significant amount of luck is required to produce our universe in the circumstances of a particular Big Bang process, I will consider this to be a major flaw.

On this basis there is a significant difference between a small/limited number of options and a vast/infinite number of configurations for reality. A small number, (even hundreds), can be worked-through on a limited timeframe. An infinite number of permutations will require an eternity to work-through, unless there is a specific process in operation that can provide a guaranteed short-cut to our reality.

The influence of Thought is pivotal here, because design and control are key ways in which short-cuts can be provided.

Physical beings may be a source of influence as we saw in Chapter 11, however they can only contribute if there was an earlier physical existence what was scattered and dynamic. Their chance of success will also be improved if there are specific mechanisms within the Universe that they can exploit.

The final point to make at this stage is that if we layer thinking beings onto a scenario (including God), then His/their use of a mechanism must not fundamentally change it, but work within its natural capabilities. This is particularly relevant when considering the actions of a God.

I feel it is important to take this approach because if we suppose that a mechanism is altered to do something else, it would represent something different and new and we couldn't assess it on the same basis as before. Our task is to assess something potentially real within the Universe, not a radical new creation which someone may deem God to invent. I have allowed a distinct row for that possibility, so our assessment of the original mechanisms can remain 'pure'.

A2.1   Assumptions

The following assumptions underpin the tables helping us to determine which combinations of mechanism and circumstance are viable while others are not:-

Column 'a' - The Total Absence of Everything assumes that there is no free floating energy; no force fields and no 'nothingness' representing the zero form of something. Equally, where relevant, it also assumes that any 'second underlying stuff of existence' is also absent initially, but that its spontaneous abilities could generate the 'stuff of Thought' as pure creation.

Across the 4 tables there is only two 'things' that might exist in the starting conditions of column 'a'. The first is a series of dimensions which represent the full extent of the void that may be filled. We occupy just 3 of those dimensions but science is seriously considering[22] that up to 11 dimensions may exist (7 hidden ones plus Time).

The second factor is God – who effectively represents a third element of reality for these purposes.

Column 'b' – this is where we find the ancient concepts of '**chaos**', **dark energy**, the potential for **fields** of influence, plus **nothingness** as the 'zero' combined state of balanced opposites such as matter and antimatter. In short, it mustn't contain anything structured or having substance.

It is also regarded as a perpetual state because if it was significantly active it is likely to be more organised resembling col 'd'. From this there are a few subtleties:-

- an environment which is almost completely formed of 'nothingness' should be regarded as being eternal (without beginning) and in a stable state, which would require something *external* to change it.

  While there could be enough free floating energy in this environment to split nothingness, the <u>perpetual</u> state I envisage would not have been able to do this by itself.

- Any level of 'movement' which did occur in Nothingness could only be at a modest level which would simply 'give it a stir' as eternal stability would have been maintained prior to the Big Bang. We can therefore ignore the raw power of any dynamic factor for these purposes.

- Nothingness that's temporary or co-exists with physical matter is dealt with in columns 'c-f'.

<u>Rows 7 & 11</u>

These have scenarios which suggest that things of substance may exist entirely within other dimensions of existence, **not** in our 3 dimensions of existence. This effectively represents a different configuration of existence and if any such material enters our reality then it would have to be transformed into our type of Matter/Energy.

My assumption is that this transformation would reduce the underlying stuff of existence into its most basic form as it enters our space, and that it would be very powerful and also able to travel far faster than the speed of light because it had not been bound by any Laws of Physics at that moment.

<u>General Factors</u>

- The presence of Intangible Thought must imply that this stuff has a small amount of power within it to influence physical matter. Even if such power was accumulated into a large resource it would not be on the scale of God and could at best tip the balance of a physical situation to enable natural forces to act, or to create some conditions where pure creation might occur.

- Thought may be dynamic even if physical matter was static. This would allow God to operate even if the physical environment was entirely static.

- If 'Nothingness' as the zero form of something exists in an environment where there is also dynamic matter then we have to regard Nothingness as a **temporary** pass-through state before it establishes matter and antimatter.

- Concentrated matter (ie. a singularity), may need a small amount of space within it to enable chemical/nuclear reactions to take place. If the singularity was infinitely dense, everything in a circle around the singularity would have to move at the same time, or movement could only occur on the surface.

- In general I have assumed that existence has to operate within the known capabilities of Matter/Energy and therefore it must follow the principles of cause and effect, and cannot do anything spontaneous.

With these in mind we can move onto the specifics.

A2.2 Physical Matter without any Influence from Thought

I will work my way down the table by row, explaining the assessment for each cell :-

Key:
- X = not possible
- HU = highly unlikely
- L/S = reliance on luck/spontaneity
- ✔ = strong possibility / Likely

Ref	Mechanism	a — Total Absence (Nothing physical in our 3 dimensions)	b — Absence of all Matter (permitting unstructured energy & forces)	c — Scattered Matter Static (eg. without Time)	d — Scattered Matter Dynamic	e — Concentrated Matter Static (eg. without Time)	f — Concentrated Matter Dynamic
1	Spontaneous Pure Creation	L/S	HU	HU	HU	L/S	L/S
2	Creation by God						
3	Energy Release - Structures			X			
4	Splitting Nothingness		S	X		X	HU
5	Induction		S	X		X	HU
6	Vacuums/Voids in space		X	X	HU	X	X
7	Self-Correcting Structures		X	X	L	X	HU
8	Chem./Nuclear - 1 explosion		X	X	HU	X	L/S
9	Chem./Nuclear Chain Reaction		X	X	L	N/a- duplication	
10	Direct Transfer - Channel/Tube		X	X	HU	X	HU
11	Direct Transfer - Dimensions	L/S	L/S	X	✔	X	L/S
12	Frameworks & Parameters		X	X	HU	X	S

Heading above table: **Starting Circumstances – Matter/Energy Alone**

Row 1 – Spontaneous Pure Creation

This considers the 'mechanism' of pure physical creation, (from absolutely nothing), which arises spontaneously, (without a cause), and *without* Thought. It is an option that's hard to ignore because it is one of only two basic possibilities for prime origin, (the other being eternal existence). The necessary assumption for this exercise is that the physical environment would be capable of doing this.

Yet many people do not agree that Matter/Energy is capable of this, and they will feel that this entire row should be coloured black. So there are differing views.

We cannot ignore the distinction between a theoretical mechanism where there is no evidence for or against its existence; and a supposed mechanism which *contradicts* all the evidence we have available. Pure creation falls into the latter category. On this basis I have to regard the evidence as a serious challenge.

Conceptually, we should know by now that spontaneity and randomness are the opposite characteristics of cause & effect and therefore contrary to the perceived abilities of Matter/Energy. Indeed, the only potential evidence that we have for pure creation is in the generation of Thought, and even this is highly disputed.

In truth, there is only theory to justify ideas about a physical environment being able to achieve pure creation and none to say it can be spontaneous. There is no evidence for major overlapping bursts of creation in the Universe and no evidence of a small

227

scale capability either, (ie. a myriad of small events doesn't seem to explain the uniform spread of galaxies or their pattern movement).

The capability for spontaneity would seem to lie with other types of stuff, however, even if we speculated that it may be possible without this, pure creation would either need to be regarded as a one-off large scale event, or something that has occurred a potentially infinite number of times in which:

- the last explosion overwhelmed everything that was there before, or
- the last explosion pushed everything that already existed beyond our view, or
- the series of Big Bangs were scattered across an infinite universe beyond our view

These add significant requirements to the basic mechanism which must challenge its likelihood. **In particular**, a one-off event makes us dependent on luck to achieve our outcome, and claiming that 'evidence of multiple events must lie beyond our ability to see it' seems a bit of a cheat... but let's be generous and say it is a half-point.

Pure creation also seems unlikely to explain the prolonged and evenly spread generation of power needed to drive the accelerating expansion of the Universe, (ie. it can't be a second blast from a 'central point'). Of course, this power might still relate to something which emerged from the Big Bang creation moment, but there is clearly a gap that needs to be filled. Another half point?

In terms of contradictory evidence there is one other factor to mention. The mathematical explanation for it requires the generation of balanced opposites, but the evidence we have is that our physical environment is almost entirely 'positive'. While we have seen clever ideas to bridge the gap from people such as Stephen Hawking and Leonard Mlodinow, suggestions that the negative elements may represent factors such as antimatter, gravity, or Thought are entirely unproven.

If pure creation did only produce positive matter, I suspect that it would only be the mathematicians who would weap.

Overall therefore, I regard the prime scenario (cell 1a) as being unlikely. Putting numbers to this, there would be one full concern and two half points.

Beyond this there are some more specific issues related to particular starting conditions. The nature of pure creation, (coming from absolutely nothing), led us to one logical possibility – that if there was an identifiable pre-condition for it to occur, this might lie in the need for a perfect void.

On this basis cells '1b-d' seem to me to be 'Highly Unlikely' because they seem unable to generate a perfect void of sufficient scale to generate an entire Universe. To my thinking it makes little difference if the starting environment was static as opposed to dynamic, because spontaneity could introduce movement anyway.

Cells 1'e,f' seem to have slightly better prospects because there would be a perfect void outside the singularity that might enable a burst of new creation. As this scenario assumes that an eternal singularity hadn't achieved anything by itself, (that is covered by 8f), it can be argued that the newly created material would be drawn into the singularity – perhaps tipping the balance somehow to cause it to explode.

Row 2 - Creation by God including Energy Release

None of the circumstances in this row are possible because this table excludes the possibility of God as an influence on background circumstances. We are considering Matter/Energy in isolation.

Row 3 - Energy Release from Structures

The key characteristic of scenarios in this row is that they provide an eternal cyclical mechanism which is capable of experimenting with different configurations of reality without being affected by the accelerating expansion of the Universe... ie. the cycle can survive. However the starting of that acceleration 5 billion years ago still has to be explained and that cannot be part of the Big Bang event itself – no matter which structures brought it about.

The hope of materialism is that the acceleration will ultimately be shown to be a natural next step in the sequence of an evolving Universe, (which means that it wouldn't need something additional or spontaneous to be generated), but this **is** currently a hope.

In terms of the core mechanism, whether we consider that there are parallel membranes/'curtains of energy' that are hanging in space beyond our field of vision, or there are any other structures that may generate a Big Bang event, we must presume that there is an interaction between more than one thing that exits. On that basis columns 'a', 'b', 'e', and 'f' are not applicable because they do not seem to permit the necessary structures. 'c' is also not possible due to the absence of any movement.

We just have to consider 3d, and this mechanism is considered different to the interactions implied by other rows dealing with Induction or Chemical/Nuclear reactions which are covered in mechanisms 5-9. It is also required to operate within our 3 dimensions, making it different to the circumstances of row 11.

The 'curtains of energy' idea does seem to resolve many problems associated with the Bang-Crunch theory as long as you accept that those curtains exist. We can even avoid the need for an Inflationary effect if we assume that such vast curtains touch across a very broad area – say 66+ billion light years. On this basis physical material would be able to spread out at speeds within our known Laws of Physics, (below the speed of light), to cover the remaining 27 billion light years distance within the 13.77 billion years available.

The emphasis therefore shifts onto explaining the accelerating expansion of the Universe. As the scenario of cell 3d works entirely within our 3 dimensions, it is hard to see how the curtains of energy could touch again within the region of our Universe 5 billion years ago, as the underlying principle seems to forbid it, (physical matter should keep the curtains apart). The outside possibility is that material in the universe had become so thinly spread 8.77 billion years after the Big Bang that the curtains came close together and gave a partial discharge across another broad area, before disappearing again.

To cover the distances involved, (85+ billion light years of the Universe at that time), the curtains would have to be get close again across an enormous area meaning that they should be within our space and visible. But we don't see them.

If it is argued that the new burst of Energy/Matter pushed the curtains out of sight again, we should also suppose that a further 5 billion years of expansion might allow them to approach again, making them visible – but no, there is no evidence for them.

While lack of evidence isn't fatal in this debate there is serious room to doubt this explanation even if it can't be ruled-out.

Alternatively we might think of our Universe as *insulation* instead of a 'physical push' that keeps the curtains apart, in which case it may be that the insulation is getting thin and ineffective. On this basis we might expect to see huge bolts of energy, (like bolts of lightning), suddenly trying to bridge the gap between the curtains.

Because our Universe operates within the Laws of Physics this would imply that such bolts of lightning could travel *no faster* than the speed of light – and this would mean that they should be visible for some 35 billion years as they crossed the vastness of space from both sides... yet, once again, we **don't** see them.

As the main conceptual purpose of this activity would be to inject energy into our universe, with an even coverage everywhere to account for the even pattern of movement, it might be argued that these notional 'bolts' had been absorbed by the material of the Universe. However bolts of lightning would miss most areas of the Universe and are therefore unlikely to achieve the desired effect. Once again, it doesn't seem to be a viable explanation unless the energy can get to all areas instantly.

I have therefore set the likelihood of cell 'd' to 'Possible' (yellow) because no credible explanation is currently available for the accelerating expansion.

Although there can be other objections to scenario 3d in relation to how such structures were put in place, and what they are actually capable of, I feel that these arguments are just part of the 'lack of evidence' issues which affect all potential mechanisms. They are not a distinguishing factor for this assessment.

On the positive side, scenarios in cell 3d <u>do</u> allow for multiple attempts at configuring existence so they would **not** be reliant on luck to produce our Universe. Such theories may also explain why we don't see the residues of previous creation events. As an example, the 'curtains' theory argues that earlier material has either dissipated to almost nothing and/or moved beyond our range of vision – (see 10.2).

Row 4 - Splitting Nothingness

The principal mechanism here is the ***splitting of nothingness*** into balanced opposites which then have to be kept apart, (the prime example being the notion of matter and antimatter). The basic scenario assumes that various things have to exist in advance of the event; from the 'zero form of something' that could be split; to the high levels of energy which may be required to achieve the split. Cell '4a' is therefore not applicable and the static environments c & e are not possible.

To avoid blurring the concepts this row does not consider that Matter/Energy has the capability to be spontaneous. It must still work entirely within the principles of causality, which means that the static environments of 'c&e' are not viable. Whether we think of this 'zero form of something' as Dark Matter or something else, it has to be regarded as an eternal part of existence if it wasn't the result of pure creation.

Columns b-f all presume that Nothingness would sit next to physical material, which has to be the result of an earlier splitting of nothingness if we are not to conclude that eternal existence had been established with a fundamental imbalance, (ie. there was more matter than antimatter). It *is* conceivable, but to my mind it's highly unlikely, so options 'b-f' would probably imply that pools of antimatter were already co-existing with patches of matter before the last Big Bang.

However it is unclear whether Nothingness as the 'zero form of something' means that this stuff exists within the other Laws of Physics, or outside any such Laws.

On the one hand, Nothingness could be seen as a 'natural intermediate point between positive and negative existence, which would suggest that it is part of an underlying structure. On the other hand we are inclined to believe that the products of any new split may have the chance to adopt a different configuration of existence in which case, for our purposes, it would be without any Laws of Physics. This is important for two reasons:

a) There is a background assumption that that once Nothingness had been split it could not be split again, which implies that an event on the scale of a Big Bang would be a one-off for our visible Universe, and therefore potentially reliant on luck to achieve our configuration.

b) If the original split of Nothingness occurred at a point in space and then cascaded outwards, it would need a process similar to the concept of 'Inflation' to cover the distances involved, and being outside the Laws of Physics may be one way to do this.

To clarify these scenarios a little more, we know from basic principles that if matter and antimatter come together again then they are likely to re-form Nothingness and so the potential exists for a further split. However from what we can see there's no evidence that matter and antimatter are re-combining anywhere in the Universe. The separation is long-term, and in most circumstances could be regarded as a one-off.

Secondly, in relation to point 'b', the pattern of movement across all galaxies still suggests that a single major event was responsible for our universe 13.77 billion years ago, which either implies that a single huge split of Nothingness occurred everywhere at once, or the initial effects spread from a 'central' point to the outer edges of the Universe over the past 13.77 billion years.

A major split across, (say), 70 billion light years of Nothingness would seem to require a means of communication and co-ordination, which in turn suggests some sort of Framework for Existence in order to avoid the need for Inflation. The 'balloon effect' may be one way to envisage this, (see below).

The alternative is to accept and explain the need for Inflation by saying that the initial state of Nothingness is outside the Laws of Physics and therefore able to break the speed of light.

In relation to point 'a' there are various ways around the need for Luck implied by a one-off split - by supposing that either

- something in the nature of matter and antimatter could **only** allow them to exist in one form, (our version of reality) – but to me this seems unlikely.

- the universe is infinite and that other major splits of nothingness have occurred beyond our field of vision, allowing nature to try out all permutations of existence – but not here..

With these potential explanations to cover some of the major issues related to the Splitting of Nothingness, one other problem remains – the need to explain the accelerating expansion of the universe, and the source of new power.

As the additional power requirement (scale of Dark Energy) calculated by NASA is 15 times the amount of physical material we can see in the Universe this is not a trivial matter. We need to find a vast new source of energy, and in this scenario the prime candidate would be to split more Nothingness.

To be able to do this would imply that residual patches of Nothingness remained after the Big Bang or that nothingness re-formed in the aftermath of that event. The accelerating expansion is very even and so we would need patches of Nothingness to exist everywhere. It would then require substantial amounts of energy to be applied to many if not all of those patches in a co-ordinated action to split them and (hopefully) release more energy than was put in.

None of the scenarios seem viable unless the split of Nothingness generates more power than it consumes in triggering the split.

With these factors in mind, let's turn to the specifics.

As column 'b' represents a perpetual state, (temporary states of nothingness would be covered by columns 'd' & 'f'), the Big Bang would be a first event in our region of space and any free floating energy/force which had previously moved through or past the Nothingness would **not** have been able to generate any significant split. This scenario therefore requires an external or spontaneous factor to change its eternal state and the circumstances cannot provide one. Without any structures (Frameworks) there is no real prospect of explaining the later acceleration through a secondary split of Nothingness. With two major issues Cell 4b is 'Unlikely' (red), however even this rating is dependent on the Universe being infinite and prompting a wider exploration of the permutations of reality to avoid luck in achieving our outcome, plus the size of the Universe being explained by one or other means mentioned.

A singularity in the traditional sense, (column 'f'), encompassing all existence, would not be possible if matter and antimatter had to be brought together, because the scenario requires them to cancel each other out and return to a zero state – removing the singularity. If you argued that a singularity (the totality of everything) was made entirely of the zero form of matter/energy then where did the separate power come from to split it? There would either have to be an imbalance in the set-up of existence on a scale that was large enough to split almost the entirety of the Universe, or there would be a need for spontaneity.

Existence would have to contain multiple singularities across the infinity of space if it was to avoid luck and there would also be a need for Nothingness to be outside all laws of physics. With multiple problems I deem 'f' - Highly Unlikely.

Yet if there were two singularities, (one positive and one negative), there would be no Nothingness, however in a sort of reverse logic, the coming together of these objects might generate a huge explosion – returning some material to 'zero form' mixed in with the remaining physical Matter/Energy and antimatter. Having broken free from their singularity status they would be free to form galaxies within the expanding Universe, however with a one-off event we would seem to be entirely dependent on luck unless infinite space was enabling an infinite number of other split experiments.

I would consider this scenario within 4d, (scattered matter), as existence would always be in more than one place, and to me it seems 'Unlikely' (red) because it fails to explain the original state and also cannot utilise the principle of inflation to explain the size of the universe because the singularities would be within the Laws of Physics.

However this isn't the only possibility for 4d. We need to consider whether a widely spread state of nothingness could split across a very wide region – say 70+ billion light years at the same moment. This could either be envisaged as one big pool of stuff exploding at the same time or conceptually as a myriad of small splits in disparate pools of this stuff.

Ordinarily the notion of a large pool blowing up would be perfectly feasible, however, given the distances involved, there would need to be a means of instant communications to achieve this, and of course we have seen this type of capability demonstrated in the Gisin[19] experiments. The differences in the concepts show themselves at this point.

We might envisage that a pool of stuff has a natural link between all elements of this 'store' because it is all directly connected, however if the Nothingness was scattered across the Universe as a myriad of discreet pools then an additional link would be required to enable the necessary co-ordination to take place.

If we take the example of 'one split per galaxy' this would represent approx. 200 billion events scattered evenly across space, and the additional links between them would effectively represent a crude Framework.

However once these links were in place a very simple mechanism may be all that's required to co-ordinate these activities. An example, the link could basically tell all parts of existence whether everything was still holding together in zero form – yes or no? If 'No' the whole thing splits, like the popping of a balloon.

In this scenario the initial split may **not** require any significant power. We may even speculate that it might be triggered by something simple – say a momentary disconnect which was interpreted to mean 'No' instead of 'Yes'.

By allowing the reaction to happen everywhere at once we have a way of avoiding the need for inflation, yet we also have to explain the later acceleration through a secondary set of splits. Regardless of the original state of affairs, any residual Nothingness would almost certainly be scattered in discreet clumps and at this point it would seem that a basic Framework would be unavoidable - representing another major requirement beyond the original mechanism. As the strongest version of this theory I feel it leaves cell 'd' as "Possible" (yellow).

However, as with cell 'b' this rating is dependent on the Universe being infinite with an infinite number of other splits occurring to explore all other permutations of reality in order to avoid luck in achieving our outcome.

Row 5 - Induction through the Action of Fields

When we consider the possibilities for Induction there is an assumption that the principles involved can operate with just the interaction of fields, and that coils of copper wire are not necessary! ☺

Induction through the interaction of fields represents a transfer of energy and possibly force using a mechanism which we are not clear about – but it is different to the opening of a channel or wormhole as envisaged in rows 10 & 11.

Induction, especially on the scale of generating a universe, is often perceived to be a one-off event in our area of space, but within an infinite universe there are possibilities for it to occur in many spread-out places.

The alternative is to assume that these interacting fields are large enough to produce a succession of Big Bangs from their own vast stocks of power, along the lines of the curtains of energy in row 3. The subtle difference is that these fields wouldn't have to physically clash, but equally there is a suspicion that these fields would only have a limited stock of energy and therefore they couldn't produce an infinite sequence of Big Bangs.

However, for me, the thing which fundamentally undermines both of the latter suggestions is that Induction works **within** the Laws of Physics, it doesn't change them, and it also requires the basic material to already exist – so it is either dependent on luck to achieve our version of reality, or it is not the prime mechanism of existence, (ie. it relies on something else to configure reality).

The main benefit of Induction is that it could potentially make a large amount of material suddenly appear at a point in space, giving the illusion of pure creation.

Once again column 'a' is impossible as forces and energy have to exist in our 3 dimensions for the mechanism to work. The process would also require movement which would exclude both of the static circumstances, (columns 'c' & 'e').

However the advantage for 5b is that these fields could exist across chaotic unstructured material, although if an eternity of chaotic activity hadn't been able to generate a Big Bang in the past (because it was still in a chaotic state), then we would need a spontaneous or **ex**ternal effect to alter that state. For this reason, plus its reliance on luck, I will show 5b as "Unlikely" (red).

While it is just about conceivable that two fields might exist within a singularity I flag '5f' as "Highly Unlikely" for a number of reasons. Firstly and most obviously, there is a need for movement with Induction, and the highly compressed nature of a singularity would make that very difficult, (especially if fields are made of particles). Secondly, the transfer of energy from one part of the singularity to another would seem to achieve very little.

Finally, Induction in this circumstance would simply be the trigger for the explosion of the singularity; the material would exist already. It wouldn't really be Induction as the mechanism, it would be the exploding singularity (covered in row 8). To be a true Induction effect would require a field to exist beyond the singularity and that the energy of the singularity transferred to a point in space elsewhere. This is not only highly unlikely but would seem to be pointless when it would be far easier for the singularity to just react itself.

As a result the principal opportunity for Induction lies in scattered, dynamic, physical matter (column 'd') - where its normal circumstances of operation would lie. To avoid the need for Inflation the interacting fields would have to be very broad – some 66+ billion light years across, and that would also require a level of intensity in that energy which could cause the formation of matter and an outward pattern of movement.

However it is difficult to avoid the earlier generic concerns about the need for luck in achieving our configuration, and there is also no explanation for the secondary and prolonged acceleration of the Universe 8.77 billion years later, (ie. we don't observe interacting fields generating lots of energy).

I am therefore inclined to flag it as 'Unlikely' (red).

Rows 6 & 7 - Vacuums in Space and Self-Balancing Structures

Although these scenarios envisage that massive amounts of energy are drawn to a point in space from all directions creating a concentration of this stuff which then explodes outwards, much of the logic that I applied for row 5 can be applied again.

The major adaptations relate to the cause of the 'vacuum effect' whether that is through some sort of pressure difference, or as a result of a self-balancing or 'mutual dependence' mechanism.

Column 'a' is ruled out in both rows because the scenarios need something to exist, while 'c' and 'e' are also ruled out because we need the environment to be dynamic and there is no source of spontaneity to activate them. 6b and 7b also suffer from the fact that they represent perpetual states of existence so events on the scale of a Big Bang could only arise with an external or spontaneous influence, which does not exist.

There are a number of additional problems which would affect 6b and 6d. Firstly it is hard to see how a void could draw-in enough material to create a Universe. Secondly, even if this was possible, the speed at which the material might flow is unlikely to be powerful enough to generate a Big Bang or to alter the configuration of existence. The reason for saying this is that nature will generally respond to a vacuum by attempting to balance the conditions in the void with the background levels of heat or pressure, not to create something that was massively in excess of the background.

In addition, there is no mechanism to explain how our version of reality came about because this mechanism works within the Laws of Physics; it doesn't change them. Finally, there is no explanation for the source of the acceleration 5 billion years ago. To this extent 6d seem to have four significant problems, making it Highly Unlikely, while 6b is not possible.

Yet in the scenario of a chain of inter-dependent universes or some other type of self-correcting mechanism, (7d), the evaporation of a universe to leave a void that could threaten the chain of existence would raise the possibility that any energy flow would cross the boundaries between different sorts of existence – particularly if they were also crossing dimensions, which might make the flows a lot stronger.

In this respect we have to assume that energy is a sort of 'universal currency' which can adapt to any environment, but a flow of material across dimensions **does** provide a conceivable way for new forms of reality to emerge, (ie. new configurations within our 3 dimensions). The transformation of material crossing these boundaries, (described in more detail at row 11), may also be a good way to explain the pattern of movement that we see in the galaxies of our Universe.

Yet there are problems with the logic. If a self balancing mechanism exists, how could it allow the environment to become so unstable that it requires extreme action? Put another way, the existence of such a mechanism should ensure that there is no need for a Big Bang event. Although the lack of evidence for a self-balancing mechanism is not a reason for downgrading its likelihood in this exercise, the reliance on another factor to make it fail does seem at least another half problem. To this we can add the other half problem of no explanation for the later acceleration of our Universe.

Finally, if there is a mutually dependent chain of existence then it may be necessary for only one type of existence to fill any gap. If so, the chain would be a fixed structure with pre-determined version s of reality within it, and without another mechanism we would have to presume a dependence on luck to put it in place, (including our specific type of reality).

With 2 challenges I therefore designate 7d as 'Unlikely'. Like 6b, 7b is "Not Possible" due to the absence of an external or spontaneous factor.

In relation to cell 6f we have to ask whether a void could be generated inside a singularity, (eg. some sort of hollow), as there is no mechanism to do this and no room for manoeuvre with such compression. I suspect most people would think it can't.

Yet the very concept of a singularity also implies that a void would have to exist everywhere around it. A serious flow from the singularity to the void would seem to invalidate the whole concept of a singularity as it would never be able to get everything in one place. I therefore mark 6f as '"Not Possible".

The odds do not improve in 7f as it's hard to see how *any* self-balancing mechanism based on the disappearance/evaporation of matter could apply to a singularity, which gathers it all together and ensures that it isn't dissipated. In short, the mechanism would seem to have nothing to replace. Yet, we don't know what such a mechanism might be, so I won't say that it is an invalid concept, just 'Highly Unlikely'.

Rows 8 & 9 - Chemical and Nuclear Reactions

These rows start from the position that Matter/Energy is eternal because there is no spontaneity and no pure creation event. 'Chemical or Nuclear Reactions' will simply occur within the material that exists, and I use this terminology to cover all natural abilities exhibited by matter/energy in our Universe including gravity, but excluding the mechanisms covered by other rows in this table.

As a result column 'a' is flagged as impossible, while the static environments of 'c & e' are flagged as unworkable because these mechanisms require a dynamic state.

Chemical/Nuclear reactions also work exclusively within the established rules of existence and seem unable to change them... other than possibly through the circumstances of a singularity, (although this is just a popular ***assumption*** as there is no concept of how the configuration of existence might change within a singularity).

Row 8, dealing with a single huge blast, is primarily confined to column 'f' because it is only a 'dynamic singularity' that could produce a single explosion large enough for the Big Bang. Indeed 8f is the basis for the Bang-Crunch theory, although we now know that this concept is dependent on an external or spontaneous influence to explain our Universe, (as our Universe would represent the last in the sequence because it cannot collapse again). Yet there is no spontaneous influence in the Materialist scenario, and there would probably be a large amount of luck in getting that spontaneity to arise ***just at the point where our Universe and life had been achieved***.

In addition there are a number of partial concerns about the mechanism from doubts about Black Holes being able to explode rather than venting their energy; to the lack of a proper explanation for the process of 'Inflation'. Many of these could be overcome if we simply deemed the mechanism to be 'capable of its tasks' because none of the other mechanisms can prove their existence either. Yet this doesn't seem entirely satisfactory because these concerns seem to contradict known facts about our reality. Equally, there is no explanation for the sudden acceleration of our universe 5 billion years ago. However even if there are ways around some the partial concerns, I would say that there are still at least two serious issues with the theory, so I designate Bang-Crunch (8f) 'Unlikely' (red).

Cell 8d suggests that a single overwhelming blast could arise from scattered matter (not a chain reaction which is covered by Row 9). Yet it has to cover the entirety of the visible universe with sufficient power to force all galaxies to adopt a uniform pattern of movement, using less material than the totality of existence, (because it is not a singularity). Even if this was possible (because we don't know what would actually be required), I can't see how this could form part of an eternal cycle within our 3 dimensions alone to explain how our configuration was achieved without luck. For these reasons I deem it "Highly Unlikely".

In general terms row 9, (dealing with spreading chain reactions), can get away with a much smaller initial reaction as long as all subsequent reactions are powerful enough to maintain the sequence. The major difficulties with this concept are in explaining

i.   how the chain reaction could spread across 96 Billion light years of distance in the time available, (as it could not travel faster than the speed of light).

ii.   how it could avoid luck in the configuration of our Universe because it doesn't seem to offer the possibility of a repeating cycle.

iii.   how the even spread of galaxies and pattern of their movement could be achieved, (ie. most distant galaxies travelling fastest, etc)..

As a chain reaction in a singularity (9e&f) would be the same as options 8e&f, these have been excluded to avoid duplication of statistics. This leaves us with 9b and 9d, and in practical terms both could have a spreading chain reaction that either

- began from a single point and spread from there, (going much faster than light)
- had multiple start points across all scattered material, (requiring co-ordination).

As chemical/nuclear reactions in these environments would have no obvious way to change the configuration of existence they must be using our configuration without offering a way to explain how this came to exist in the first place. As a result, both seem to leave us dependent on luck, (ie. leaving problem 'ii' unresolved).

Problem 'iii' is also challenging because a chain reaction would be likely to set everything moving at the same speed, not at variable speeds dependent on distance.

While multiple start points do offer a way to cover the distances involved in a way that a single start point cannot, there is still no explanation of the acceleration which began 5 billion years ago. Although this could be a second round of these co-ordinated myriad events, (as we don't see a spreading wave rippling through the Universe), this is a very unspecific answer, describing a different effect to the original Big Bang.

Overall I would argue that there are at least 2 major challenges, so 9d becomes 'Unlikely' with a reliance on luck. Yet 9b suffers from other key issues:-

- that the Matter/Energy in this scenario is believed to be more thinly spread than scattered matter (c&d), and therefore incapable of sustaining a chain reaction of any size, (ie it couldn't be expected to spread very far).
- as this chaotic environment would have been stable for all eternity, a spontaneous influence would have been needed to change it;

On this basis, I would say that 9b isn't possible.

Rows 10 & 11 - Direct Transfer of Matter with/without Transformation

The basic scenario is that enough material to form our Universe was transferred from other parts of an infinite existence to the centre point of our Universe where it then exploded outwards in all directions as the Big Bang we popularly imagine.

Ordinarily a direct transfer of matter/energy from one part of existence to another along a channel in space would require 'stuff' to exist, and in the circumstances of Prime Origin this would also mean it had no beginning and was in dynamic form, invalidating cols 'a, c, & e' – especially in Row 10.

However the circumstances of row 11, which operate across hidden dimensions of existence, opens-up another possibility in relation to column 'a'. The environment might be completely empty in our 3 dimensions, but very full in others. A hole that opened-up between dimensions could allow a massive amount of 'stuff' to transfer into our level of existence, generating enormous power in two ways: firstly by the speed of transfer, and secondly by the *transformation* of matter from one state of existence to another, momentarily taking it out of all Laws of Physics and thereby possibly explaining 'Inflation'.

To understand some of the arguments which will follow in later paragraphs, we first need to picture what might happen in these various circumstances.

Row 10 envisages that all existence and movement occur within our 3 dimensions and must adhere to the Laws of Physics. In these conditions we know that any tube, 'sea current', or channel, will essentially require material to flow in one direction from 'A' to 'B'. While we have to ask what might cause material to flow like this and to stay within the confines of a particular path, we must also appreciate that on the scale of a universe the distances that would be travelled by that Matter/Energy are enormous.

We could potentially watch a long line of material floating down this very long route, possibly appearing as a visible 'thread' across the Universe when we observe the cosmos with our telescopes. So much material would be required that this line of stuff would be very long. In all probability, even travelling at the speed of light we would need to see many of these threads if an entire universe was to be moved on any realistic timeframe. Yet our reality check is that we **don't** see such threads.

Of course, this may have occurred in the past, not now, and a stream of energy might not be visible, however the material then has to explode outwards to form our Universe, and within row 10 it would have to stay within the Laws of Physics, so there is no way for it to cover the 96 billion light years width of our Universe today.

However there is another popular concept known as a 'wormhole' amongst sci-fi enthusiasts. This suggests that the fabric of space can be bent into a circle so that the two ends of the tube/channel end up sitting right next to each other, even though the original points being connected were very distant – possibly even on 'the other side' of the Universe. The concept also suggests that the tube would shorten as a result, to just cover the tiny gap that now exists between the ends, and not the original distance through the Universe. Visibly, if you looked into such a tube it might seem very short.

As a means of instant travel across vast distances you can see why such concepts have been used in many science fiction storylines, but that doesn't make them real, and nor do the mathematical equations which have been offered to support the notion.

The fact that this mechanism would not only require the bending of space but the bending of an existing Universe, is often skated-over. To my thinking the power requirement would be beyond belief... but faith is a wonderful thing.

Another suggestion therefore is that a tube or channel across the Universe might generate special circumstances within it to accelerate matter/energy well beyond the speed of light.... yet a tube or channel generally isn't special, and the only thing that I have read which would make it special would be converting it to a wormhole... which may not be possible anyway.

While you may guess that I remain sceptical about these notions within row 10, (our 3 dimensions), a lack of evidence is not a factor in this assessment. I have to assume that a wormhole may be possible.

So in principle, **visually**, at the 'despatch end', physical matter would be sucked towards the entrance where it would be drawn-in. We don't have any idea what force would actually drive this 'suction effect' or the flow down the channel, however it would be a necessary part of the mechanism which we again have to accept as part of the concept. Some sort of pressure difference can be implied.

It is in the next steps that we begin to see flaws. The material which emerged at the 'delivery end' of the tube/channel would be travelling in the direction of flow at that point, like water from a hose. The same would probably be true of a wormhole. Yet that is **not** the pattern of movement which is associated with the Big Bang. Here material must travel outwards in **all** directions not just one – like an explosion.

Yet the scenario of Row 11 **does** provide the 'correct' pattern of distribution as stuff emerged from a 'rip', 'hole', or 'channel' through to another dimension.

If there are other dimensions of existence then something tangible may have the same co-ordinates as a patch of empty space in our Universe, but the extra co-ordinates for the additional dimensions means that it would not appear in our space. This can be likened to two pieces of paper, one laying on top of another, where the co-ordinates in 2 dimensions are the same, and their location only differs in the 3rd dimension, height.

It means that the distances which might be travelled during this transfer would be very short, and might also occur across a very broad region of space – helping to explain the size of our Universe.

A rip in the fabric of space between those invisible environments and our own 3 dimensions would be like a hole which you could see from any angle, but would seem to go nowhere. Equally, any material passing through it wouldn't have any sense of direction in relation to our 3 dimensions, so stuff coming out of the hole would be able to travel in any direction, not just one – resembling an exploding singularity.

Crossing dimensions would almost certainly change the fundamental configuration of matter and reduce it to its most basic form. That, in turn, may also increase the power of the flow because, during transformation, it would momentarily be free from all Laws of Physics – potentially allowing it to travel many times faster than the speed of light – an explanation for Inflation? This scenario applies to 11'a/d/f'.

In contrast, **10**a is 'Not Possible' because there could be **no** transfer of material into our 3 dimensions so there would be nothing to create a flow or current.

The opening-up of tubes or channels also implies a degree of structure that is absent in scenario 10b, and as this is a perpetual state it would require a spontaneous or **ex**ternal factor to make it viable, which this scenario doesn't provide. It is 'Not Possible'.

In relation to 10d, it is hard to envisage that flows from a pipe could mimic the Big Bang. Not only do we have the problem of 'direction of travel' but also that the material being moved would also remain within the existing Laws of Physics – meaning that there was no explanation of how our configuration was achieved. There would also be no explanation for 'Universe Inflation' to account for the size of the Universe. The only conceivable way in which I can make this work is to suppose that many tubes/channels existed, all transferring material to a particular point in space with sufficient force that when they arrived at the same place and collided, they generated the conditions necessary for a Big Bang. In many ways this would be equivalent to the build up of a singularity but with the added complication of tubes or channels.

While 10d isn't intended to create a singularity my main objection to this idea is that there is nothing to explain why all these notional channels/tubes should all point to the same location. I simply don't believe that this is realistic/possible.

Finally, in relation to cell 10f, I feel that the scenario of a singularity is 'Highly Unlikely' because a tube or channel in our 3 dimensions requires a beginning and an end, and there would be nowhere other than the same singularity to go to.

While 'crossing dimensions' offers a lot of potential, I categorise 11a as 'Unlikely' because the absence of any previous material indicates that this might have been the first time that this occurred, which would deny nature the opportunity to experiment with different configurations of reality – making us dependent on luck. As a first occurrence after an eternity of existence, this would also require a spontaneous or **ex**ternal factor to introduce the necessary level of change, (a reason for the transfer which this scenario **doesn't** provide). The same spontaneous/**ex**ternal influence may also be needed to explain the sudden acceleration of the universe 5 billion years ago.

All of these factors also apply to 11b, as its perpetual state of chaos would mean that this would be the 'first hole' between dimensions.

Cells 11d and 11f offer much better prospects. They each need a way for nature to experiment with different configurations of reality in order to guarantee our type of Universe, and the two main ways to do this were discussed in Chapter 10: a singularity, (11f), and clashing pools/curtains of energy, (11d).

However if we are to exploit the potential of crossing dimensions in order to explain the process of inflation then both of the original mechanisms need to be adapted. Taking the latter first, the notion of the 'curtains' changes, so that they are instead envisaged as vast pools of energy which are now held in separate dimensions, kept apart by the dense physical material in our 3 dimensions.

As in row 3, when our Universe spreads out too far these curtains/dimensions might touch across a very broad area, (say 70 billion light years across), to account for the size of our Universe within our known Laws of Physics, or alternately a crossing of dimensions might provide a way to explain Inflation, (because material would be freed from all physical laws as it was transformed when entering a new set of dimensions).

11f returns us to the idea of a singularity, which might be able to transfer all the material needed for a Universe very quickly into our space, while also explaining inflation through a crossing of dimensions. This concept also suggests that a singularity might be the only thing with sufficient power to break through to the other dimensions. We can then modify the Bang-Crunch cycle so that each time a singularity forms it will punch a hole through to other realm and move itself across, generating a Big Bang event within those dimensions.

Once the crunch happened again and a singularity re-formed, a new rip would occur possibly allowing the material to move back to the original dimensions – and this would establish an eternal cycle of activity which would allow nature to explore every permutation of existence.

Both 11d and 11f have still to explain the sudden acceleration in the expansion of our Universe beginning 5 billion years ago, and there is added significance for any explanation within 11f because the cycle would become broken by this process, potentially making us dependent on luck to achieve our outcome – unlike 11d.

On a positive note, the acceleration requires a vast new source of energy and one of the few sources available may again be the hidden stores of energy in other dimensions. We might therefore envisage a weaker form of interaction with the 'curtains' of 11d. However this seems a **less** likely possibility in the circumstances of 11f where the entirety of a singularity would have drawn all material out of those other dimension at the last Big Bang, **and** there would be no explanation for any new rip or channel. When assessing 11f I felt that these two points were linked - amounting to one factor - a lack of explanation for the acceleration.

Unlike row 3, 11d dangles the possibility that the thinning universe allowed the curtains to 'come close' in other dimensions... yet the whole concept of parallel dimensions means that they should always be close. It is possibly better to re-phrase this as the physical material in our universe acting as insulation, preventing energy from jumping between them. Could it be that after 8.7 billion years the insulation was getting porous – allowing a partial flow of energy across the entirety of the universe? This more gentle flow could also arise across the entirety of the visible Universe.

Given the amount of energy that seems to be required, this is the only suggestion I can think of for explaining the acceleration within the Materialist viewpoint, (and in some way or other there does have to be an explanation). Of course, that explanation doesn't have to come from the Materialist viewpoint.

On this basis I categorize 11d as a 'Strong Possibility' (blue), and 11f 'Unlikely' (red) due to the lack of explanation for the acceleration and a need for luck due to the ending of the eternal cycle.

Row 12 - Configuration Settings for Existence

This scenario presumes that dynamic Matter/Energy has a mechanism by which new parameters of existence can be set, communicated, and applied across the whole Universe. In practise this would require something structured, (perhaps a Framework for Existence), and so columns 'a', 'b', 'c', and 'e' are discounted.

The initial concept of a Framework was to enforce a standard configuration across all of existence, but this seemed to rely on luck to achieve our outcome if there was only ever one configuration on offer. However if we are to allow nature to experiment then there has to be a source of new settings. In terms of speculation this may either be

a) by reference to a particular point in the Universe which generates new settings,

b) by getting feedback from a dynamic universe which may occasionally generate new settings in some unknown way, but also in a manner that prompts the framework to use those settings instead forcing them to change back into the original format.

c) through the actions of a singularity which somehow manages to generate new settings and then uses the framework to standardise those settings everywhere.

In addition, new settings should only arise very rarely to achieve a stable existence, (implying that the different settings would have to be compatible with each other).

In one sense we can assume that only those settings which are compatible/viable would survive, however if some 'random setting generator' was constantly churning these things out, there could never be any stability in the Universe. Something must determine that a new long-term standard will not be changed without 'good reason'.

We each have to decide whether this mechanism of generating and assessing new configurations is part of the Framework or something separate to it. To me it seems like something conceptually different, making it a challenge to the process.

As we can assume that the implementation of new settings through the Framework would be achieved everywhere at once, thanks to instant communications across the Universe, it could be applicable in 12d – effectively creating a myriad of transforming explosions across a spread-out Universe. The main difficulty would then be to create the pattern of movement that we see across all galaxies which would presumably occur as a result of interacting pressure waves or a distortion in the fabric of space. Of course this places additional mechanistic requirements on the theory – and to my mind a second layer of concern for 12d. Finally, both 12d and 12f fail to explain the sudden acceleration in the expansion of our universe 8.7 billion years after the Big Bang.

For these reasons I categorize 12d as 'Highly Unlikely' while 12f is 'Unlikely'.

A2.3   Mechanisms for Origin working with Physical Thought

The results from the analysis *above* show how likely the different mechanisms would be in producing our Universe with the basic properties of Matter/Energy, (excluding Physical Thought), yet these results will change if we layer-on additional capabilities.

We have seen that there are various types of Thought which might exist, related to the different types of stuff from which it might arise. If you are a materialist then you still have to accept that Physical Thought, (as I label it), **is possible** through physical beings, although it would be based entirely in Matter/Energy and therefore bound by the rules of single cause: single effect. So this analysis will look specifically at how Physical Thought could improve the chances of success in each set of circumstances.

The potential for Physical Thought is undeniable because we are here, and if nature has been experimenting with versions of reality through a series of Big Bang events across eternity, then such beings will be virtually guaranteed at some point prior to the last Big Bang... and quite possibly at some point before our civilization emerged within the past 13.77 billion years.

Materialists have to believe that life will inevitably emerge if the circumstances are right, and it also seems reasonable to suppose that a sophisticated civilization would not only explore how the Universe works, but would also use its knowledge – say if it believed that it faced annihilation at the next Big Bang, and had nothing to lose.

In this sort of context we can begin to narrow the circumstances where physical beings might be able to operate. Firstly, they could only exist in a dynamic physical environment containing scattered Matter/Energy, (ie. a Universe similar to our own, not a singularity or in the midst of raw chaos). These physical beings could not be present where conditions were too extreme and unable to support physical life, and by the same token they would have limited physical power (they would not be Gods), so they would have to utilise the mechanisms that nature had already developed.

Overall, this means that they can only survive in the circumstances of column 'd' and mainly operate on rows 3 to 12. Yet as a slight caveat to this, the potential exists for physical beings to influence events that may destroy them by taking action in advance. These actions would therefore occur in the conditions of column 'd' but could influence a mechanism that was part of column 'f'. While I preserve this separation in the table below, I have merged the 2 columns in the summary table for ease.

The presumption is that once a civilization crosses a threshold of ability, its scientific progress will be very rapid in comparison to the lifespan of a Big Bang cycle, and therefore it would have plenty of time to not only develop very powerful capabilities, but also to plan their use, (by generally manipulating the inputs to a process).

		Starting Circumstances with Physical Thought					
		a	b	c	d	e	f
X = not possible				Scattered Matter			
HU = highly unlikely					Dynamic		
L/S = reliance on luck/spontaneity							
✔ = strong possibility / Likely							
Ref	Mechanism						
1	Spontaneous Pure Creation				▓		
2	Creation by God				█		
3	Energy Release - Structures						
4	Splitting Nothingness						
5	Induction				L		
6	Vacuums/Voids in space				HU		
7	Self-Correcting Structures				L		
8	Chem./Nuclear - 1 explosion				HU		▓
9	Chem./Nuclear Chain Reaction				L		
10	Direct Transfer - Channel/Tube				HU		
11	Direct Transfer - Dimensions				✔		
12	Frameworks & Parameters						

In tinkering with such fundamental processes they **wouldn't** be able to use spontaneity or randomness, but could be acting out of sync with the rest of physical existence, thereby representing an **ex**ternal factor for these purposes.

However we also have to recognise that their limited knowledge could potentially lead to catastrophic errors which may prematurely end the eternal cycle of existence, (leaving it in a state without life). Conceptually, this negative impact may outweigh any benefits that such a civilization might bring – so we each have to judge how much risk they would represent. One way of thinking is to say that because such events would have happened in the past, they didn't make a mistake by leaving us in this state.

Row 1 – if we believe that it is possible for Gods to engineer the right circumstances to enable pure creation to take place, then physical beings may also be able to do so, although to a lesser degree. The generation of a perfect void may be one example where physical beings could act in this capacity. They may even survive a blast of pure creation on the scale of a Big Bang if the void was created remotely, (at some distance). However we have to ask what would be achieved by it from our perspective?

These sophisticated beings would already exist in an environment that was capable of supporting life, and pure creation leaves some uncertainty about the outcome which they probably wouldn't be able to immediately configure. It therefore seems a very risky strategy if the environment that we ended-up with couldn't support life.

If this civilization was unable to design and guarantee a configuration outcome then strategically their act of pure creation would be no different to any other act by nature itself ... except if their void produced a lesser burst of creation over a very wide area that was able to boost the energy levels in the Universe – say 5 billion years ago.

On balance I am tempted to say that their presence improves the chances of row 1 slightly by helping to explain the accelerating expansion, so I will raise it to Unlikely.

Row 2 - unchanged as God remains outside the scope of this scenario.

Row 3 – Structures such as the 'Curtains of Energy'

Whether or not 'curtains of energy' actually exist, the theory suggests that some types of structure might be able to create an eternal cycle of Big Bang events while being immune to the accelerating expansion of the Universe. In this sense nothing would seem to be achieved by increasing or reducing the power of the blast. It also seems highly unlikely that any civilization would be able to influence the curtains of energy to force them into a designer output.

Yet the accelerating expansion of the Universe which began some 8.7 billion years after the main event is a reality that still needs to be explained, even if we can't see the purpose in it. Physical beings **could** provide that missing factor if they were able to generate the massive amounts of energy required on a Universal basis... and that's the difficulty. There's no apparent way for the curtains of energy to be activated in the way that we need part-way through the cycle.

Even if an advanced civilisation discovered the curtains of energy and tried to tap into them, then realistically they are only going to produce localised effects and not the even injection of massive amounts of energy that would seem to be required across the **entire** Universe.

Other hidden dimensions of existence, (row 11), or a Framework for Existence, (row 12), are not present in this row even if we were to argue that some level of thoughtful co-ordination might be possible. The only possibility is that some sort of chemical/nuclear process would kick-in to convert matter into energy as a new source of power to generate that acceleration over a sustained period of time.

As explained more fully in Row 8 the idea that Dark Matter might be utilised by a civilization for this purpose is undermined because NASA doesn't believe that there is enough Dark Matter to account for the Dark Energy requirement they have calculated.

As the potential for physical beings to intervene is highly complex, highly risky, and with significant residual problems, I leave my original assessment unchanged: 'Possible' (yellow).

Row 4 – Splitting Nothingness.

The very detailed analysis in A2.2 showed that the splitting of Nothingness might be 'Possible' (yellow) in cell 4d under one specific scenario. This envisaged that a huge pool of Nothingness would initially split across a very broad area (70+ billion light years) at the same moment. A secondary set of splits would then be required to generate the power for the acceleration, and to achieve a similar level of co-ordination across fragmented pools of stuff there would need to be a framework.

While we are all familiar with the potential for a big store of volatile stuff to explode at the same moment, the situation here is different because of the distances involved.

Our question now is whether the presence of physical beings could improve the situation at all?

To achieve sophisticated life forms without luck would either mean that an infinite existence had been exploring all forms of reality in a series of 'split' experiments across unlimited space, or that nothingness could **only** produce our form of Matter/ Antimatter. These **are** possibilities but we have to ask ourselves how likely they are.

Once physical beings do exist then we might speculate that they would be able to identify and remotely trigger the splitting of the separate patches of Nothingness needed for the acceleration effect.

If there was a natural Framework then they might be able to exploit it, however this doesn't seem to improve the likelihood because we envisaged this mechanism anyway.

The additional benefit would therefore be if such beings **constructed** the necessary framework artificially, using technology. Not only do I find this very hard to believe, it would also imply that Nothingness wasn't set up to respond to a simple instruction – so it is likely to require a vast amount of power to generate that myriad of new splits.

Personally, I feel that this scenario stretches the imagination beyond reasonableness, especially as the acceleration is not a one-off moment but a continuing process and we do not currently see splits occurring everywhere.

I have therefore left the categorization unchanged – 'Possible' (yellow).

Rows 5-7 – for very similar reasons to row 3, (particularly the inability to generate sufficient power to influence Big Bang events), these assessments also remain unchanged. The provision of a design and the application of some targeted power wouldn't change any of the fundamental circumstances. The natural environment would be operating on a scale far bigger than they could ever hope to influence.

To be clear, unlike Row 1, the power of any void that physical beings might create would be based on the relevant 'pressure difference' that existed within the environment nearby. Even if those pressures were there, it is likely that this civilization would have to engineer a huge void in order to move Matter/Energy on a large enough scale to create a Big Bang.

In relation to Induction, the likely size of the fields needed to generate a Big Bang is too large for them to realistically manipulate. The destabilising of a self-balancing mechanism on the scale of a Universe would seem equally impossible.

As a result, the main opportunity for living beings is not in producing a designer Universe but in explaining the acceleration which started to happen 5 billion years ago. But in the context of this row, that is a relatively minor point.

I therefore leave the rankings unchanged.

Row 8 – Chemical/Nuclear Single Blast

The Bang-Crunch cycle has been proposed as the way nature might achieve our configuration of Universe using a chemical/nuclear mechanism, without resorting to luck. Living beings would be able to emerge in the period of existence between singularities if conditions were sufficient to support life.

Although this wouldn't necessarily be our version of reality, and different forms of existence could produce stronger or weaker chemical/nuclear reactions than we are used-to, it is still believed that Matter/Energy would operate **within** its version of the Laws of Physics and would not change them.

As far as we can tell, a singularity, (representing the totality of a Universe), would always be made of the same stuff in the same form, and would presumably need to remain unchanged if an eternal cycle were to be preserved, so it is unclear how it could change the configuration of existence each time it blew. Without chance or a spontaneous/random influence we can only surmise that this 'inevitable' change would relate to the 'chemical composition' of the emerging singularity, (before it was complete). The window of opportunity for any civilization to design a 'poison pill' that could influence the formation of a new reality, lies in that build-up period.

Although the 'poison pill' approach is conceivable, it would almost certainly result in the destruction of that civilization. To guarantee a suitable outcome (from our perspective) their attempt to produce our Universe after they were gone would require them to have near perfect knowledge. Yet they **couldn't** have experience of any previous exploding singularity as their existence would only have emerged after the previous Big Bang.

The potential for them to get things wrong, and to end the cycle prematurely, would seem to cancel out the benefit of them trying to accelerate nature's own experiments.

Yet that assumes that their deliberate intervention would want to change existence.

The alternative is to suggest that their intervention would be to **preserve** what already existed by breaking the cycle - say accelerating its expansion beyond the point that gravity could make it collapse again, 5 billion years ago.

By implication this approach would mean that *their* configuration of reality was the same as *ours* which allows the previously successful Bang-Crunch theory to explain how this version of existence came about before they, (as an external influence following a distinct path to Matter/Energy), broke the eternal cycle.

This also overcomes another more subtle factor in the explanation of our origin, (that I mentioned in Chapter 11): the eternal cycle should only end after our configuration of existence was achieved; or at least after producing a configuration that could support life, which just happened to be ours. If physical beings broke the cycle (accidentally or on purpose), it would be a great way to ensure that this happened at the right time.

Yet this theory still leaves us with the problem being considered by many scientists – how chemical/nuclear processes operating within our Laws of Physics could find a **new large source of energy** and co-ordinate its release across the entire Universe.

The potential for instant communications across the Universe, (per the Gisin experiments), offers an outline method for control, but unlike rows 3 & 4 there is no source of energy to tap into that's sufficiently large. (There isn't enough Dark Energy in existence to cover NASA's estimates for Dark Energy).

Equally, for similar reasons, I do not see any way for a civilisation to influence any stuff which may surround our universe. The acceleration remains unexplained.

As the involvement of Physical Beings doesn't seem to provide any additional explanation for 'Universe Inflation' either, to explain the size of our Universe, I have chosen to leave the previous rating for 8d as 'Highly Unlikely' and 8f as 'Unlikely', although physical beings might remove the need for Luck or Spontaneity.

Row 9 - Chemical/Nuclear Chain Reaction

If we exclude the possibility of a singularity or a Framework for Existence then there is no apparent way for chemical/nuclear processes to change the fundamental configuration of the Universe where material is already widely spread-out. This would mean that the environment before any large scale chain reaction was in exactly the same configuration as our reality using the same Laws of Physics.

Aside from there being no explanation for this configuration, there is also little prospect of an eternal cycle emerging that was based on chain reactions.

The main opportunity for living beings to influence events is therefore as an **external** influence to trigger the accelerating expansion of the Universe 5 billion years ago.

The process described for Row 8 above using Dark Matter is the most likely approach and would be very similar to the chain reaction being proposed in 9d, however in this case there is no imminent prospect of a singularity forming, so there is no apparent reason why any civilization in the circumstances would attempt to do such a thing.

As physical beings do not offer any additional explanation for

- how the speed of galaxies would increase the further away they are from us, or
- how our configuration of reality was achieved without luck

I have decided to leave the rating unchanged as 'Unlikely'.

Row 10 - Direct Transfer - Channel/Tube

Firstly, any transfer of material within our 3dimensions, would leave Matter/Energy in the same configuration of existence, so this mechanism doesn't provide any explanation for our version of reality.

It also seems highly unlikely that physical beings would be able to generate the amount of power required to move an entire Universe even if they were to create a wormhole. Getting all scattered material to a wormhole, or transferring it at the speed of light down a normal channel or tube would be a very slow process. Opening a wormhole next to a singularity also seems pointless. If they wanted to create a Universe why wouldn't they do it where the material already existed?

So if the only contribution of living beings was to draw additional energy into our universe from elsewhere – perhaps as a new source of Dark Energy to account for the sudden acceleration it is hard to see how this could be done practically using tubes or channels because the energy would have to be distributed evenly across the universe.

For these reasons the involvement of physical beings doesn't change my rating.

Row 11 - Direct Transfer - Dimensions

This is the other scenario where physical beings who can only exist in the circumstances of column 'd' might still be able to influence column 'f' by *preventing* the formation of a singularity to preserve the status quo.

The idea of 'singularities crossing dimensions' (11f) explains many things but it still suffers from the fact that its cycle would have ended, making it dependent on luck to achieve our outcome. Yet it can be argued that if physical beings are the only factor which could end the otherwise eternal cycle, they would only be able to do so after they cycle had achieved it prime purpose – their creation – which must be acceptable.

So the prime questions are whether they *would* be the only factor able to break the cycle, and whether physical beings could account for the accelerating expansion of the Universe which began 8.7 billion years after the Big Bang event, (5 billion years ago).

To be honest, there is no way of knowing whether physical beings are the only way to end the cycle of activity. However if there were other ways (such as spontaneous or random activity), then the cycle could never have been expected to be eternal in the first place. So there are reasons why living beings might be the only end to an eternal cycle but the logic is not conclusive.

In rows 3 and 8 I rejected the idea that physical beings could generate the level of energy required for the acceleration within our 3 dimensions, but if they were able to access huge amounts of energy by generating another rip/channel between dimensions the idea begins to seem more likely. Much would depend on whether these beings had the ability to generate such a breach between dimensions.

In relation to 11f we have previously assumed that it was the distinctive power of a singularity that was the only thing powerful enough to generate a rip between dimensions. On that scale or requirement I can't believe that it would be possible for any civilization to achieve a transfer of material from other dimensions of existence.

However we simply don't know what might be possible so there may be other ways to generate such a rip. If correct we must also recognize that the injection of energy from other dimensions must be spread evenly across the entire universe in the time available.

On the basis that crossing dimensions may allow energy to travel at speeds far faster than the speed of light, this might be achievable. For me this is the key difference with 3d and 8d where the hidden pools of energy cannot be accessed in a suitable way.

Yet there are still issues with the notion that 11f could find a sufficiently large pool of energy after the previous singularity had moved everything out of the other dimensions.

Although the circumstances of 11f have been strengthened, doubts remain, so I have upgraded 11f to 'Possible' (yellow). The presence of physical beings doesn't seem to worsen the prospects for 11d (and may even improve them), so I leave it as a 'Strong Possibility' (blue).

Row 12 - Frameworks & Parameters

If a Framework for Existence is ever discovered then it seems likely that physical beings would find a way to exploit it. Physical beings could be the source of new designer settings which was a major problem with this scenario.

Yet we have to explain how physical beings could initially arise, and unless we were totally reliant on luck to generate a suitable configuration of existence in the first place, an alternate means of generating configuration settings would still be required.

I am not persuaded that physical beings could be the source of a Framework which would either have to surround or be embedded in every particle of existence. This would be very hard to achieve once matter/energy is scattered.

If 12d represented the intermediate state between two singularities in the Bang-Crunch cycle, then physical beings could be the means by which the cycle was ended but if their chosen means of achieving this was through new parameters in a Framework for Existence they would probably be signing their own death warrants.

Either through accidental actions by a 'mad scientist' (per the scenario in chapter 11), or from a sense of the greater good, etc. there are ways to argue why they might do such a thing, but to be honest, that is not our concern. If they did take such action to produce a 'designer Universe' then it would fulfil a purpose in explaining our origin, although strategically this would only accelerate the process that nature was undergoing anyway.

The bigger danger is that their attempts to force a change halted any cycle of natural experimentation as well as leaving an environment that was unable to support life.

However instead of trying to change the configuration of existence they may simply wish to use the Framework as a way to accelerate the expansion of the Universe by (say) getting dark matter to transform into energy. If a Framework could allow them to trigger certain effects while preserving the basic configuration of existence then they may indeed be the missing factor to explain this aspect of reality – yet it would still not explain how nature could experiment with different configurations of existence by only implementing stable and viable forms of existence.

As it would also seem strange that use of a Framework produced a phased release of energy to explain the acceleration, the main basis for increasing the rating to 'Possible' is down to them being a source of new configuration settings.

The table below shows how the viability of different mechanisms changes with the added influence of a second type of underlying stuff, which brings a range of new capabilities with it. In basic terms it would put true spontaneity & randomness into the mix. In more advanced circumstances it may provide a degree of awareness or a sense of direction to cover the action of basic particles, and finally in the most sophisticated circumstances it may add fully developed Pluralistic or Intangible Thought.

Yet a more direct influence might lie in the ability to convert this other substance into physical energy. If this were possible, then we envisage the other stuff to be less powerful than Matter/Energy, but we don't know how much of it there might be, so it could produce significant effects. However I don't want to confuse it with the notion of God who has unlimited powers. If this analysis is to be valuable we need to position it as contributing **less** power and ability than God.

		Starting Conditions with Intangible Thought					
		a	b	c	d	e	f
		Total Absence	Absence of all Matter	Scattered Matter		Concentrated Matter	
		Nothing physical & no Thought whatsoever	permitting unstructured energy & forces	Static (eg. without Time)	Dynamic	Static eg.without Time)	Dynamic
Ref	**Mechanism**						
1	Spontaneous Pure Creation	**L**	**L**	**L**	**L**	**L**	**L**
2	Creation by God						
3	Energy Release - Structures			✓	✓		
4	Splitting Nothingness		X	✓	✓		
5	Induction		**L**	**L**	**L**	HU	HU
6	Vacuums/Voids in space		X	HU	HU	X	X
7	Self-Correcting Structures		X	**L**	**L**	HU	HU
8	Chem./Nuclear - 1 explosion		X	HU	HU	**L**	**L**
9	Chem./Nuclear Chain Reaction		X	**L**	**L**	N/a- duplication	
10	Direct Transfer - Channel/Tube		HU	HU	HU	HU	HU
11	Direct Transfer - Dimensions	**L**	**L**	✓	✓	✓	✓
12	Frameworks & Parameters		X	✓	✓	✓	✓

Legend:
- X = not possible
- HU = highly unlikely
- L/S = reliance on luck/spontaneity
- ✓ = strong possibility / likely

Once again, this analysis builds on the assessments of previous tables as it adds an extra layer of capability. Unlike physical beings, an additional type of stuff would not need special conditions in which to operate so it will potentially apply to all columns in our table. However where scenarios imply the existence of physical beings using Pluralistic Thought, column 'd' will again be the main focus.

In terms of achievements, the capability to be spontaneous or random can do a number of things. It may for instance enable an eternal cycle to be changed, and/or allow a static environment to become dynamic. For this reason columns 'c' and 'e' now become copies of their partner columns 'd' and 'f'. Each pair could therefore be merged, but for clarity of comparison I will leave all columns showing at this point and will only merge them in my overall summary table as a way of saving space.

I will work my way down the Rows, focusing on changes from the previous tables.

Row 1 – Spontaneous Pure Creation

If you believe that spontaneous pure creation has occurred it is necessary to believe that the environment is capable of acting in this way, a 2nd type of stuff which is capable of spontaneous acts would remove a major issue because it would demonstrate that certain capabilities were possible. Those abilities raise the ratings for 1a-d by one level, and also enables the Universe to begin from a 'blank sheet'. Yet a burst of new creation alongside a singularity would seem to offer little benefit with or without Thought, so I have not changed my assessment of 1e & 1f.

If we believe that a 2nd type of stuff exists then each substance can have a separate origin; meaning that there are different ways to view cell 1a. The starting position for 1a prohibits anything from existing in our 3D space, however if a 2nd type of stuff has the ability to be spontaneous, we can either argue that

i.   it spontaneously created itself (meaning that it has the power of pure creation too).

ii. the background environment is the thing which has these capabilities and that the 2nd substance would be tapping into those capabilities if it enabled spontaneity.

The implication from 'i' is that the 2nd type of substance would be the first to be created, and it is highly likely that it could only generate intangible material like itself plus some form of energy. There would need to be a secondary process to generate physical matter, but more importantly perhaps, there is the potential for this 2nd substance to evolve and develop more sophisticated capabilities before it generates physical matter, and therefore able to give a sense of direction if nothing else.

Yet option 'ii' has no particular leaning, so the background environment could potentially generate either type of substance. (The 'fact' that only one type would exploit the potential of spontaneity is not a clear pointer in this respect).

The physical emptiness of the environment would mean that cell 1a was the first creation event so the emergence of any substance is likely to be in its most crude/raw form. Matter/Energy could be evolving alongside the underlying stuff of Thought so if we are to avoid a dependency on luck to achieve our configuration of Universe, we

1) must prove that only one type of Universe was possible, which is unlikely, or

2) the chaotic evolution of Matter/Energy would take so long that it would allow the other stuff to develop more sophisticated abilities that could provide crude direction

3) the first burst of creation produced a myriad of Universes across infinite existence, in which case ours was bound to be one of them.

Overall, many of the old uncertainties remain about 1a, however the presence of other stuff removes one of them allowing it to become 'possible' (yellow).

Cells '1c&e' become viable because spontaneity can make the environment dynamic.

Cells '1b-f' all consider that some material has existed forever, and that pure creation happens alongside it. It is here that we find the potential for Intangible Thought, and possibly Pluralistic Beings in cells 1c&1d.

The issue here is the degree to which these manifestations of the other stuff could influence physical reality. In order to distinguish these circumstances from those of God I have assumed that the physical power of such stuff is quite limited, and it therefore either has to work in subtle ways, or it needs an enormous volume of this stuff to directly change a small amount of Matter/Energy.

Across all of the above logic, including the pace of evolution, there seems more opportunity for Intangible Thought to influence chaotic material than structured material, because structures would probably require greater levels of influence to change them. Cell 1b therefore has an opportunity for shaping/design if more of the 2nd stuff is generated through pure creation, (although I don't see this as full control).

The potential of Pluralistic Beings to influence their environment with more direct power is countered by the fact that physical beings would first need the physical environment to adopt a strong and stable structure. Whether intangible Thought increases the chances of physical life is a moot point.

The half-way-house comes in speculating whether Intangible Thought could fashion the fabric of space and/or a Framework for Existence. These notions tend to imply quite sophisticated mechanisms based on concepts and communications which unthinking Matter/Energy shouldn't need; the development of messenger particles, and instant communications over large distances being examples.

If the Fabric of Space implies that it is not made of the same stuff as Matter/Energy, (which needs it to be in place before Matter/Energy can occupy that space), then this may be a way for Intangible Thought to enhance its powers. However this is pure speculation, and concepts of an 'Internal Framework' may make this less likely.

The emergence of physical beings might offer a way to trigger the accelerating expansion of the Universe 5 billion years ago, but to my mind in the context of spontaneous pure creation this doesn't significantly improve the chances of success.

However across these different ideas, there are 'glimmers of hope' so I have raised my assessment of 1'c&d' to 'Unlikely' (red).

Row 2 - Creation by God

This scenario is unchanged as God remains outside our scope.

Row 3 – Energy Release from Structures like the 'Curtains of Energy'

As the 'curtains of energy' are generally portrayed as only being able to produce a burst of Matter/Energy, (not a 2nd type of stuff), there is no means by which this other stuff could be generated. It would therefore have to represent an eternal presence in the same way that the curtains were also eternal.

In the earlier assessments cell 3d was able to explain the origin and size of our configuration of the Universe without resorting to luck, by imagining that the curtains touched across a broad area, (70+ billion light years). This explanation avoided the need for Inflation, however it struggled to explain the sudden acceleration in the expansion of our Universe.

There simply wasn't a sufficiently large power source that was spread evenly across the Universe to account for this effect. However if we now suppose that a 2nd type of stuff does exist then this problem might be overcome, especially if this stuff was able to spontaneously create itself, gradually replacing what was used for physical purposes.

We know that, if it exists, Intangible Thought would have the ability to influence physical effects, (as indicated when our minds influence our bodies). Could it be that vast quantities of the underlying stuff could achieve much bigger physical effects?

This might be envisaged in two main ways:

1) Although the underlying stuff of Thought is often perceived to be physically weaker than Matter/Energy, we don't know how much of this substance existed in the past, so it is conceivable that it might be enough to generate the equivalent amount of power as Dark Energy which NASA has calculated.

2) We have speculated that the curtains of energy may have been able to interact again once the insulating effect of the Big Bang was diluted over the first 8.7 billion years, leading to a more modest discharge of energy. However we also speculated that it might discharge in a similar manner to bolts of lightning, (which would be too much of a localised effect). So perhaps the influence of the 2nd stuff would be sufficient to spread the power of these powerful discharges, to achieve the even expansion we seem to observe.

Although highly speculative, there is a reasonable logic to each of these suggestions so the question I have asked myself is whether my residual doubts are enough to prevent 3d from becoming a strong possibility?

It is part of this scenario that a 2nd type of stuff exists. We have to accept that as a given. What we don't know is what that substance is truly capable of. We my simply deem it to be powerful enough to achieve either 1 or 2 above, in which case 3d would turn blue, (along with 3c which could also be made dynamic by this stuff).

On the other hand, we can look at the detail of these ideas to see if they work at a practical level. For me there are two main issues. Firstly any discharge of physical energy would be working within the Laws of Physics as we know them because this energy would not be crossing dimensional boundaries. It would therefore be travelling no faster than the speed of light.

A discharge between the curtains of energy would imply that they had come much closer together (within our line of sight), making those discharges relatively short-lived, (which is why we don't see them), otherwise those bursts wouldn't be able to cover the distances involved in the time available. As we don't see those curtains, I struggle to believe that this chain of activity left no trace with which to verify that it occurred, so in combination these factors leave significant doubt about option 2 above.

However if you accept the basic premise that there was enough power available for option 1, it would not be affected by this thinking because a co-ordinated discharge by the 2nd substance across the whole universe would affect everywhere immediately.

As the 2nd substance is deemed to have Spontaneous/Random capabilities it can explain any features of the explanation that require them. We can also say that if these influences are confined to generating the acceleration effect, there is no implication of luck in achieving our configuration of reality as this would be down to the cycle of clashing curtains, (especially as the cycle would not break by the acceleration).

In this discreet way, through option 1, cell 3c&d do become fully viable, (blue).

Row 4 – Splitting Nothingness into Balanced Opposites

As a mechanism, the splitting of nothingness is generally associated with physical Matter/Energy and not any other type of stuff that underpins existence, however it is possible that the 'Stuff of Thought' may be the balancing factor to physical Matter/Energy and therefore created as part of the split – bringing the opposite properties to cause & effect.

Without this, the 2nd type of stuff would have to be an eternal feature of existence sitting alongside Nothingness, (pure creation not being a factor within this scenario).

An eternal presence for this other substance would always provide its additional basic influences, however an eternity might allow more sophisticated capabilities to emerge, (even Intangible Thought), by the time of the Big Bang. Bet if the 2nd stuff only emerged as part of the split, then it could only provide basic influences.

In one sense, the capability to be spontaneous may allow cell '4b' to alter its perpetual state of chaos to generate a Big Bang. Yet by the same token, the presence of spontaneity should make 'eternal 4b' impossible – which is my inclination.

The basic circumstances of 4f mean that it is still only viable if the singularity was made entirely of the 'zero form of something', however, where the earlier difficulty had been in finding a source of power to generate the initial split, it is possible that the 2nd substance might generate enough power to do this and if it had developed intermediate or high level abilities as well, then it might do so with a greater sense of purpose rather than as a spontaneous/random effect.

To avoid luck in generating our version of reality we would again be requiring an infinite existence to be generating all other forms of reality elsewhere, or requiring only our version of Matter/Antimatter to emerge from the splits. Nothingness would also have to be outside the Laws of Physics in order to generate an Inflationary effect.

I have previously accepted that one of these additional factors may be reasonably accepted however I feel that requiring both of them does raise a valid challenge.

This brings us to the point that also has to be explained for 4d – how to generate the power necessary to explain the accelerating expansion of the Universe?

As we saw earlier the actions of physical or intangible beings to power and co-ordinate a second set of splits artificially, (using technology), seems unrealistic given the vastness of space. However there is now one other possibility.

If a 2nd type of stuff was present in significant quantities everywhere in the Universe and could be converted into physical energy it may either

1) be sufficient to directly power the acceleration, or

2) provide the energy with which to triggers further splits in pools of Nothingness across the Universe.

Option 2 still requires the identification of those pools of Nothingness as well as a means of co-ordination, which again seems to fall back onto living beings or the natural emergence of a Framework.

However option 1 has clear potential and therefore I set the likelihood of 4e&f to 'Possible' (yellow) and raise 4c&d to a 'Strong Possibility' (blue).

Row 5 - Induction

Without an obvious source for a 2nd type of stuff we would have to assume that it was an eternal presence. On the same logic as Row 4 there is a dilemma about whether the perpetual chaotic state in 5b was possible in the presence of a spontaneous influence, however even if this could be resolved it would still mean that any Big Bang event would be a first instance and subject to luck in achieving our outcome.

There is still no obvious way for a process of Induction to be part of a regular cycle of events that would allow nature to experiment with all versions of reality. This is because Induction works within the existing Laws of Physics and would not change them. The ability to be spontaneous doesn't change this unless it also enabled pure creation which is **not** a part of this scenario.

For this reason, unlike Row 4, I feel that both 5b and 5d will remain unchanged.

As If there *was* a perceived way for Intangible Thought to influence the shaping of that energy burst then it would increase the likelihood of this mechanism, but there isn't, and it doesn't have the power to force a change.

While physical beings might provide an explanation for the accelerating expansion of the Universe it leaves the question of the size/movement of the Universe unresolved.

Rows 6 & 7 – Voids and Self-Correcting Structures

As the ability to be random/spontaneous doesn't seem to affect column 'b', and neither voids nor self-balancing mechanisms on the scale of a Big Bang would be affected by the presence of Intangible Thought, all original assessments remain unchanged.

Row 8 – Chemical /Nuclear Single Explosion

When we looked at the potential for physical beings using Physical Thought to influence the formation of our Universe through the use of a 'poison pill', they only seemed to have a marginal impact because of the risk that they might make a mistake and because they would only accelerate the process of change that nature was undertaking anyway through Bang-Crunch. As Intangible Thought wouldn't have the power to correct a mistake or try again, these basics don't change.

Intangible or Pluralistic Thought would still be prone to error as this would either come from insufficient knowledge, or the fact that background spontaneity or randomness, (coming from this alternate stuff), might disrupt planned events as well.

As an academic point the ability of nature to be spontaneous might potentially explain how a singularity would be able to change the configuration of existence within a chemical/nuclear process – although this was assumed to be possible anyway.

However, the main potential is for this 2nd substance to inject additional energy into an existing scenario. This offers a way for 8f to explain the accelerating expansion of the Universe but if it ended the cycle before life had been achieved then it would remain dependent on Luck. I feel that it is taking this scenario too close to the idea of God if we suggest that it could make its move on a planned basis.

I therefore raise the potential of 8e&f to 'Possible' (yellow) but leave 8d unchanged because scattered matter is not believed to be capable of generating a single explosion on the scale of Big Bang.

Row 9 – Chemical /Nuclear Chain Reaction

Without a singularity, chemical/nuclear processes operating on scattered material do not seem able to change the configuration of existence and the presence of an alternate type of stuff doesn't seem to change this. On this basis there is also no realistic prospect of a cycle of activity to test out each permutation of existence and therefore the origin of our existence remains unexplained and potentially dependent on luck

The potential influence on 9d comes from physical beings influencing the pattern of movement amongst galaxies – especially by engineering the accelerating expansion of the universe, however I still cannot see how they could access enough energy within the confines of our 3 dimensions.

Their influence doesn't seem to be hugely significant in the context of this scenario so the rating remains unchanged at 'Unlikely' (red).

Rows 10 & 11 – Transfers

The presence of a 2nd substance may provide the ability to undertake spontaneous actions, but not spontaneous pure creation, (which is the remit of row 1).

The main factors that we have to consider are

- the influence of living beings, (without the power of God), and
- whether the physical power that this 2nd stuff might unleash would be sufficient to make a difference on rows 10 & 11.

As any transfer of material within our 3 dimensions, (row 10), would be travelling below the speed of light and within discreet channels it remains difficult to see how it could generate the even pattern of expansion in the Universe we observe today. Row 10 therefore seems to be unaffected by the potential of a 2nd substance.

The main opportunities for row 11 again seem to lie in the ability to explain the accelerating expansion of our Universe, perhaps in combination with generating a hole/channel through to another dimension.

As 11d remains a 'Strong Possibility' the main focus lies in the impact on 11f.

If the direct conversion of this 2nd substance into physical power provided enough energy to power the acceleration directly then there would be enough justification to raise 11f to a 'Strong Possibility' (blue).

If not, then the additional potential must lie in the ability of this stuff to generate another rip or channel through to the other dimensions.(possibly with the assistance of living beings and their technology. Within this scenario the residual challenges that I identified earlier are:

a) A need to explain how a new rip/channel might be created through to another dimension in the absence of a singularity.

b) Whether enough energy might exist in the other dimensions to generate the acceleration effect 8.7 billion years after the last Big Bang, given that the previous singularly should have taken almost all of the material out of that other environment.

My earlier logic shows that there are significant doubts about these issues, so the strongest option lies with the influence of this 2nd stuff as a direct source of power.

Row 12 – New Parameters for a Framework of Existence

Without a source for a 2nd type of stuff, we have to assume that it is eternal and may therefore have developed more sophisticated forms – including Intangible Thought. If it exists, this stuff could be applied across all starting scenarios not just cell 'd' so there's potential for it to influence the parameters being implemented by a Framework.

It is unclear whether Intangible Thought could help the formation of a Framework. While Frameworks are generally perceived as being based in Matter/Energy because that is what they would control, they also seem to utilise 'concepts', because those who believe in such things will generally consider the settings to be non-physical things; coded instructions which would rely on instant communications around the Universe. So a Framework may be more likely if a 2nd type of stuff was to exist.

Intangible Thought can be used at any stage where such a Framework exists – including immediately before or even during a Big Bang. This is because Intangible Thought doesn't need a lot of physical power in this context. That would come from the Framework itself. Thought would only have to insert new parameters into the Framework, whose physical power would implement the change.

Spontaneity or randomness might produce a change of configuration but it would do so without being allied to any particular cycle and might introduce incompatible and destructive changes if there is no assessment prior to implementation. So although the concept of 'intangible beings' will seem 'off the scale' to some people, they form a good model for what we would hope-for in a generator of new parameters.

Unlike physical beings, Intangible Thought/beings wouldn't be destroyed by fundamental changes to the configuration of Physical Existence so if intangible beings got something wrong, they could presumably try again quite easily.

Against the background of this thinking, I would **not** expect a Framework to be present in either 12a or 12b, so these gradings remain unchanged.

In the context of scattered material, (12c&d), there is still the question of how the required pattern of movement might be achieved by a change of configuration, but if this essentially returned all physical matter into raw energy again we have seen that there are ways in which this could potentially be done. The added benefit of a 2nd type of stuff would be the generation of the new parameters that triggered the change.

Intangible Thought can raise the rating above that achieved by Physical Beings because it largely eliminates the destructive potential for making mistakes.

12c/d/e/f are raised to a 'Strong Possibility' (blue).

A2.5 Mechanisms of Origin Influenced by God

The table below shows how the viability of different mechanisms changes with the influence of a dynamic 'Creator God' working with His environment.

Different religions around the world have different concepts of God's role in relation to the prime origin/creation of our existence, with some of the major religions even playing to **both** of the main possibilities: creation vs transformation.

For our purposes I will consider an abstract form of creator God whose own origin will boil-down to 4 main possibilities: the different combinations of either eternal existence or pure creation, that is either within our realm of existence, or outside it.

If God exists then He must be made of something, and this will either be something unique to Him or it will form part of a broader environment. Indeed, if this stuff is unique to Him then some might say that He forms the entirety of that particular environment, however if God interacts with our Universe then this must be part of a broader existence of which He is a part.

For this reason I personally prefer to say that any God would be part of the totality of existence but He may be made of entirely different stuff to us. On this basis and for the purposes of analysis & clarity, I will refer to 2 possible types of deity:

  a) One that is part of our existence and it therefore made of the same types of stuff,

  b) One who is **not** a part of our existence and made of something entirely different.

Both will be capable of wielding unlimited physical and mental power.

The subtly different impacts which these alternate types of deity might have, become apparent when we compare different ideas about the origin of our environment with the origin of God. This is best illustrated by running through the main possibilities :-

  • If God is eternal but our environment, (the bits we are made of and can access) had a beginning then, with the greatest respect, God would be '**Type b**' not 'Type a'.

  • If God and our environment have the same origin, (whether eternal or with a beginning), then God may either be 'Type a' or 'Type b'.

  • If our environment was eternal and God had a beginning, then it is most likely that He would emerge from and be made of one or other of the same types of stuff – '**Type a**'.

Hopefully you can see the potential interplays between the different permutations of prime origin. So my second assumption is that if you are made of something then your capabilities cannot exceed the capabilities of that stuff. Put another way, the capabilities of a 'Type a' deity will be subtly different to those of a 'Type b' deity.

As we saw in Chapter 11, 'pure creation' by definition comes from absolutely nothing, however God may create the right circumstances for pure creation to occur. Yet, if this capability exists in our environment it is also possible for pure creation to occur without God. From this logic, Row 1 considers the possibilities of God utilising the natural mechanisms that exist in our environment.

Row 2 reflects creation on the command of God, to generate things that wouldn't happen naturally. This could simply be the result of His design where He uses the capabilities of our environment in a unique way. Yet a 'Type b' God may also have capabilities that are beyond our environment, to generate something radically different.

This means that God has the capability to do anything He wants in row 2, but if He chooses to utilise the mechanisms in the other rows, He will work **within** their capabilities and only manipulate some of the circumstances around them or the products that emerge from them. In this way we are able to assess the mechanisms for what they are, rather than the concepts being blurred to represent something else.

To be clear, if God were to change a natural mechanism (to operate in a different way and therefore to have different capabilities), He would have created something new, and this would be part of the assessment in row 2.

In relation to all rows other than row 2, and in broad terms, God may exert His influence to change each scenario by being able to do 3 things:-

- He can act as a spontaneous trigger in the absence of any other mechanism, and

- He can be a source of design that could allow Him to manipulate the starting conditions for some mechanisms to guarantee an end result - avoiding luck, and

- He can be a direct source of power and influence to shape the things going into a mechanism, but leaving the mechanism itself untouched.

X = not possible
HU = highly unlikely
L/S = reliance on luck/spontaneity
✓ = strong possibility / likely

		Starting Conditions with the Influence of God					
		a	b	c	d	e	f
		Total Absence	Absence of all Matter	Scattered Matter		Concentrated Matter	
		Nothing physical & no Thought whatsoever	permitting unstructured energy & forces	Static (eg.without Time)	Dynamic	Static (eg.without Time)	Dynamic
Ref	Mechanism						
1	Spontaneous Pure Creation	✓	✓	✓	✓	✓	✓
2	Creation by God		✓	✓	✓	✓	✓
3	Energy Release - Structures			✓	✓		
4	Splitting Nothingness		✓	✓	✓	✓	✓
5	Induction		L	L	L	HU	HU
6	Vacuums/Voids in space		L	L	L	X	X
7	Self-Correcting Structures					X	X
8	Chem./Nuclear - 1 explosion		HU			✓	✓
9	Chem./Nuclear Chain Reaction		L	L	L	N/a- duplication	
10	Direct Transfer - Channel/Tube		HU	HU	HU	HU	HU
11	Direct Transfer - Dimensions	L	✓	✓	✓	✓	✓
12	Frameworks & Parameters		X	✓	✓	✓	✓

Row 1 – Spontaneous Pure Creation

The cells in this row could only operate if the capability exists naturally in the environment. A command by God would be covered in row 2.

In my earlier assessment I identified 3 main issues with the concept of spontaneous pure creation. The first represents concerns that pure creation may not work in the intended manner with Matter/Energy alone, however that is now resolved with the presence of another type of stuff which provides the capability for spontaneity.

The second issue was based on speculation that the only viable way to generate the pattern of movement in our Universe was through a single major explosion and not a myriad of smaller events. On this basis and for various reasons there was a concern that only one major creation event may be possible, (with only much smaller creation events occurring afterwards). This would make us reliant on luck to achieve our configuration of Universe. However the influence which a deity could bring to shape the outcome would now guarantee our configuration.

Remember - the main purpose of spontaneous pure creation is to generate more stuff, (increase the pot of existence). It can then be shaped by other forces, including God.

The final issue relates to the ability of different circumstances to create the right conditions for pure creation – especially in relation to the scale of a Big Bang event. One of the traditional things which God could do is to engineer those pre-conditions, allowing nature to then take its course from that start point.

With the 3 objections now satisfied there is just one residual question – could God exist in all of the circumstances – particularly cell '1a'? The answer is yes, but in different ways depending on the type of deity we envisage. There are 3 possibilities:

- A 'Type a' deity was **not** present at 1a but the environment is naturally capable of generating pure creation and will do so regardless of God's presence.

- A 'Type a' deity **was** present in 1a but only after an initial burst of pure creation which brought Him into existence, but nothing else.

- There was a 'Type b' God but nothing else at cell 1a, and therefore his influence could be brought to bear straight away.

All cells in row 1 are therefore fully viable options.

Row 2 - Creation by God including Energy Release

As already mentioned, the core principle behind this row is that God can directly shape any aspect of existence to create something new, and having different capabilities to what would happen naturally. This would include generating new things by pure creation at His simple command.

Logic suggests that God can only have powers which match the stuff that He is made of, so if pure creation at His command doesn't reflect any ability within our existence He must be made of different stuff, (Type b).

I have assumed that the existence of God implies that at least one other type of stuff does exist beyond Matter/Energy and, on row 1, that *our* environment has the capability of pure creation.

However the circumstances of cell '2a' mean that a 'Type a' deity could not exist because there would be a complete absence of Matter/Energy or Intangible Thought, so the viability of this cell depends on which type of deity we envisage.

A 'Type b' deity might exist in these conditions and could make the cell fully viable, but a 'Type a' deity could only kick-in if there had been an earlier burst of spontaneous pure creation - but there would now be a caveat to the basic scenario and a dependency on something else happening in addition to God's will. For this reason I categorize 2a as 'Possible' reflecting the different possibilities.

The other cells in this row would all be fully viable, although it is worth clarifying that regardless of 'Type', if God sought to manipulate the 'scattered material' of columns 'b,c&d' He would have the spontaneity & pro-active influence to trigger such actions plus the power/ability to specify our outcome both in terms of configuration & pattern of movement.

Row 3 - Energy Release from Structures

As 3d is already deemed a 'Strong Possibility' but based on a rather tentative scenario, the influence of God can only strengthen that possibility. In addition, with an ability to be spontaneous, either God or the natural environment could generate movement in the static conditions of 3c, so this cell remains indistinguishable from 3d.

The focus therefore turns to the other cells to see if God can improve the chances of success with His influence. Sadly, I don't think He can.

I have previously designated 3a, 3b, 3e, and 3f as invalid because those conditions couldn't provide the necessary structures, (eg. curtains of energy), and God's influence *doesn't* change this. While God may well be able to put those structures in place they would **not** be 'natural' and would therefore be covered by Row 2.

Row 4 - Splitting Nothingness

In setting the context for this section I feel it is important to re-state that earlier ratings were bolstered by the expectation that ways could be found to guarantee that our version of reality emerged from a one-off split in our region of space.

There were 2 conceptual ways in which to do this, which I why I accepted the possibility as part of the prime scenario. The first was to assume that the splitting of Nothingness could only produce our version of Matter/Antimatter, while the other approach assumed that an infinite existence with an infinite supply of Nothingness would be able to experiment with different versions of reality by sparking different Big Bang events in different parts of existence.

Unless one of these applied we would be reliant on luck to achieve our version of reality. In these circumstances it is interesting to note that God would seem to offer the only way to guarantee our outcome from a one-off split in Nothingness.

As God knows everything we may speculate that He would understand how different configurations of existence might arise from the 'splitting' process, and therefore He would be able to configure any inputs to that process to guarantee that the desired outcome would be achieved.

Column 'a' remains 'not applicable' because there would be nothing for God to split. ('Pure Creation' to fill the gap is covered by Rows 1 & 2, and other dimensions are covered by Row 11).

Two key limitations of cell 4b have been: i) the need for a spontaneous or **ex**ternal influence to break unstructured chaos out of its eternal malaise, and ii) the concern that there would be insufficient power to sustain a chain reaction long enough for a full universe to emerge. God could now provide the impetus for change plus any extra power that was necessary to trigger a split in the available Nothingness. In short, 'b' becomes viable. The robustness of the earlier ratings is therefore bolstered.

In relation to columns 'e & f' the basic logic remains true, that the only realistic starting condition is a singularity composed entirely of the 'zero form of something'. This became a viable possibility when we envisaged that a 2nd type of stuff might be converted into physical energy in order to trigger the split.

The residual difficulties were in guaranteeing that our version of reality would emerge, and that an Inflationary effect would arise because the Nothingness was outside the Laws of Physics. While ways were available to potentially overcome these points I have previously only been prepared to modify the scenario for one of them not both. However with God being able to resolve the luck factor, I feel that 4e & f can now be raised to a 'Strong Possibility' (blue).

On this basis, with the exception of 'a', all cells seem fully viable.

Rows 5 -7 Induction, Vacuums, and Self-Balancing Structures

These are all mechanisms where energy/matter might migrate to a particular point in space as a result of some sort of imbalance or 'pressure difference'. As before, God's spontaneity and power should enable Him to change the starting conditions of columns 'b', 'c', and 'e' to make them dynamic and viable, however Column 'a' would remain invalid as there would be nothing to generate the effects.

In the context of the 3 mechanisms, the basic circumstances of column 'f' (relating to singularities) do not seem to be affected by the presence of God because there would be very little for Him to manipulate to achieve a flow that moved away from the singularity. Even with a chain of Universes, flows would normally be to replenish material which had drifted away – which is not the situation of a singularity.

So our focus should turn to my previous assessment of column 'd'.

If God is to utilise the natural mechanisms of the Universe then He must either
- 'tip the balance' of activity that was happening anyway, or
- engage in moving huge masses of material to manipulate the starting conditions.

If God can only use His influence to manipulate starting conditions, then in the circumstances of 5d – Induction in scattered material, we can imagine that His influence would be used to intensify the fields and build up the power that was available to be transferred. Yet in overview this would seem to be a pointless exercise. If the mechanisms themselves just moved the stuff of existence with minimal transformation, what would be gained? He could just place those resources in the correct position directly.

In the circumstances of 6d, God would be moving all Matter/Energy away from a location in space to cause a vacuum of sufficient power to create a Big Bang – and that could be a very large volume indeed. In addition, God would have to keep all flows out of the zone where He was creating the vacuum, and if that wasn't a quick process then it would make the task all the more difficult. So it is again hard to see what the benefit would be.

The reasons why He might choose to use a mechanism is not really important to the assessment of viability and the fact that it is conceivable for a Creator God to take actions like these must increase the theoretical chances of success.

262

It may also be true that God wouldn't need to undertake massive movements, but simply tip a balance somewhere. We simply don't know what the circumstances might be. However if we regard the different mechanisms as 'labour-saving devices' then 'Tipping the balance' would probably require a lot less effort/power and may therefore represent a stronger possibility.

Because cells 5d and 6d presume that Matter/Energy was already in the correct configuration, there is still a question over how our configuration of reality came to exist without luck. While God may provide an answer to this as part of an earlier process, it would mean that 5d and 6d would not represent the true point of origin.

In addition, even if God manipulated the starting conditions for such mechanisms, they would be based within the Laws of Physics and would be unable to cover the 96 Billion light year distance in the time available, and there would seem to be no special circumstances to cause an 'Inflation' effect.

Overall, for these reasons, I have raised the likelihood of 5d and 6d to 'Unlikely' as they still have 2 significant issues.

However the circumstances of 7d are different because the transformation of material across dimensions, or other different configurations of reality, do provide unusual circumstances in which an Inflation effect might arise and where the configuration of existence might be manipulated.

While God may be capable of dissipating a Universe to trigger a sudden self-balancing effect, it may equally be true that a different and more subtle sort of imbalance might be created by Him across dimensions, tipping a balance while triggering a Big Bang scale of event. I judge 7d to become 'Possible'.

Rows 8 & 9 - Chemical/ Nuclear Reactions

These rows consider how God might use chemical/nuclear reactions to create our Universe, thereby utilising the eternal physical material in His environment.

It also builds on the possibilities that might arise from a 2nd type of stuff producing living beings with Pluralistic or Intangible Thought. God may influence them as well.

If God wanted to achieve our particular Universe and then stop at that point while only working with existing mechanisms, (not creating something new), then the Bang-Crunch cycle now becomes fully viable because He has the ability to maintain the cycle until it produces a sufficiently advanced civilization and then guide the actions of living beings by making them aware of relevant factors.

Put another way, if there is the potential for natural occurrences to break the eternal cycle He must intervene to prevent them from happening until the civilization arises, and if He is **not** able to tinker with the mechanism in this row, He must manipulate the circumstances which feed it.

Equally, by making the beings aware of relevant factors, and hopefully not tinkering with their minds, their interventions either using a 'poison pill' or by manipulating Dark Energy are now likely to be successful. In short, 8e&f become fully viable, because living beings provide Him with a method of control and timing within an existing creation that emerges from the Bang-Crunch cycle based on a singularity.

Yet I feel that 8b,c&d can only be 'Possible' at best, because even if God injects a huge amount of energy to generate the single blast required for these scenarios, He couldn't make them part of an eternal cycle without directly changing the 'mechanism', and therefore there is no way to guarantee our outcome without a direct configuration (which would be row 2). On the positive side, initiating a huge blast everywhere at once across widely spread-out material may account for the size of our Universe and, through a second injection of energy, may also explain the acceleration.

The need for an eternal cycle, even for God, within these circumstances is due to a bizarre factor. God will be able to predict outcomes which run on processes based in cause & effect, but I can't see how even He could anticipate an event which is truly spontaneous or random, and the potential for this would exist if there was a 2nd substance underpinning existence. If He is to avoid direct creation, and leaves the natural mechanisms untouched, an eternal cycle would seem necessary.

The serious issue which must be faced is if spontaneous activity stopped the Bang-Crunch cycle (8f) prematurely. If this happened, then God would presumably have no other choice than to re-start the cycle, however in some people's eyes this would no longer be the normal mechanism but an example of direct creation (Row 2).

Counter arguments can be made in the following ways:

Firstly, the ability of a spontaneous act to suddenly stop the cycle is not guaranteed, and it becomes even less of a possibility when there is no build-up after the change has been made. The only way to stop God from intervening is if the spontaneous act occurred during a Big Bang moment, and even then, it would have to occur before life was generated for it to be significant. (Examples might be if the power of the Big Bang was increased beyond the ability of gravity to hold it back; or indeed, if a second source of energy kicked-in to accelerate the expansion).

Across eternity even the most unlikely possibilities may become inevitable, but in this case timing is important so a catastrophe is not guaranteed.

The second point is that if He simply re-started the mechanism He would not have invented something new, (it would still be the same process **not** a new creation that worked in a different way).

I am happy to leave 8f as a 'Strong Possibility'.

Yet I feel that 8a remains Not Applicable because if He were to inject all of the stuff of existence into our 'blank' environment, this **would** amount to direct creation, (Row 2), and there would be no other sources to populate our environment.

In relation to Row 9 and the possibilities for a spreading chain reaction in cell 9d, the only way for the size of our Universe to be covered in the time available would be for Him to trigger multiple chain reactions across the Universe.

Yet the problems with this lie in i) explaining how our configuration of existence was achieved without luck because there is no known basis for an eternal cycle based on this mechanism, and ii) how the pattern of movement came about in our Universe.

Through careful timing in His triggering of the mechanism, (which would not change it), God may have been able to achieve the pattern of movement by ensuring that the individual explosions were close enough together to cause the pressure waves to press against each other and then force everything to move outwards. Unfortunately, there is no mechanism to configure our specific version of reality within this scenario, and any direct configuration by God would fall into row 2. For this reason 9b,c&d can be no better than a Possibility requiring luck.

Rows 10 & 11 - Direct Transfers within Our Reality & Across Other Dimensions

Compared to the previous assessment with Intangible Thought, God's main influence over any transfers of material would be on row 11 where it crossed dimensions and would therefore have to be transformed. (God couldn't make the mechanisms of row 10 viable without changing them into something else, (which would be covered in row 2).

As cells 11c-f are already blue, attention turns to 11a & b where the circumstances indicate that there had been no prior Big Bang activity and on previous opportunity to configure existence. It requires a source of spontaneity to break the previously eternal circumstances, as well as a way to guarantee our outcome from a one-off event.

While God can easily act as a source of spontaneity, His critical involvement would be to ensure that the correct configuration of existence arose by manipulating the inputs to the process. (I feel that directly manipulating any stuff that emerged within our level of existence from other dimensions would amount to direct creation. He could do it but it would be part of row 2).

The comfort is that we envisage all stuff crossing the dimensions would be reduced to its most basic state before being set in a new configuration, and so the cause & effect factors which determined particular outcome could only relate to very basic issues that could be influenced by the inputs and the nature of the dimensions.

In relation to 11a, much depends on the type of God which was present (Type a or b).

As a reminder, if your concept of God is 'Type a' then He would be made of the same type of stuff as us – within our 3 dimensions. In other words, if our 3 dimensions had absolutely nothing in it before the cross-over from the other dimensions, then such a deity would not have the ability to influence any of the events.

God could only have an impact on events if He was made of different stuff, ('Type b'), and probably able to cross dimensions. As we don't have clarity on what God is I will say that the maximum potential for this cell is 'Possible' (yellow).

The potential for cell 11b is slightly different because this envisages that some chaotic stuff (without any Laws of Physics) might exist in our space before additional material crossed-over from other dimensions. While Matter/Energy might have no configuration it is possible that 'the stuff of Thought' may have evolved into God, presenting the outside possibility that He might be able to influence events.

This would imply that over an eternity of existence without physical matter, God would somehow have learned enough about the nature of existence to predict and influence outcomes on any transfer of material from other dimensions.

On this basis, both types of God would have the potential to influence outcomes and guarantee our version of reality from a one-off event, and on this basis I classify 11b as a 'Strong Possibility' (blue).

Row 12 – New Parameters for a Framework of Existence

With a Framework for Existence the generation & implementation of new configuration settings would be easy for God, and would guarantee our Universe as an outcome.

If applied to a singularity, 12e&f, a radical change of configuration might explain the effect of Inflation, allowing a single explosion to account for all of the observed effects in our Universe.

The implementation of new settings for existence on an environment with scattered material would require God to account for the pattern of movement in the Universe that we observe. As in Row 11 His ability to do this may simply reflect good timing for His intervention to trigger the effect.

Re: 12a&b the lack of a framework still means that this method cannot be used.

A2.6 Summary Overview

This summary of the findings recognises that with Physical beings, only one column is affected, and that with both the underlying stuff of Thought and the influence of God, the outcomes of static environments would be the same as their dynamic counterparts, so only columns d &|f are shown.

Legend:
- X = not possible
- HU = highly unlikely
- L/S = reliance on luck/spontaneity
- ✦ = strong possibility / Likely

Ref	Mechanism	Total Absence – a (Nothing in our 3D)	Absence of Matter – b (chaos in energy & forces)	Scattered Matter Static – c	Scattered Matter Dynamic – d	Concentrated Matter Static – e	Concentrated Matter Dynamic – f	With physical beings – d	2nd type Underlying Stuff & Intangible Thought – a	b	c/d	e/f	With God – a	b/c/d	e/f
1	Spontaneous Pure Creation	L/S	HU	HU	HU	L/S	L/S		L	L	L	L	✦	✦	✦
2	Creation by God		HU	X							✦		✦	✦	✦
3	Energy Release - Structures		S	X							✦		✦	✦	
4	Splitting Nothingness		S	X		X	HU			X	L	✦	✦	✦	✦
5	Induction		X	X	HU	X	HU	L		L	L	HU		L	HU
6	Vacuums/Voids in space		X	X	HU	X	X	HU		X	HU	X		L	X
7	Self-Correcting Structures		X	X	L	X	HU	L		X	L	HU			X
8	Chem./Nuclear - 1 explosion		X	X	HU	X	L/S			X	HU	L			X
9	Chem./Nuclear Chain Reaction		X	X	L	N/a- dupli'n		L		X	L	N/a		L	N/a
10	Direct Transfer - Channel/Tube		X	X	✦	X	HU	HU		X	HU	HU			✦
11	Direct Transfer - Dimensions	L/S	L/S	X	HU	X	L/S	✦	L	L	HU	HU	✦	HU	HU
12	Frameworks & Parameters		X	X	HU	X	S			X	✦	✦		HU	HU

267

Notes

1	NASA's WMAP programme (Wilkinson Microwave Anisotropy Probe – launched 2001), mapped the density of the Cosmic Microwave Background radiation in the Universe across 5 frequencies. This allowed scientists to remove 'foreground contamination' in a number of ways and then estimate the CMB's temperature via Planck's Law. It was then used to measure the geometry of the Universe and consider its evolution – particularly in relation to cosmic inflation theory (which did seem to be supported by the findings). The results also concluded that 95% of the early Universe was composed of Dark Matter plus Energy and that the physical universe emerged 400 million years after the Big Bang. "Nine-Year Wilkinson Microwave Anisotropy Probe (WMAP) Observations: Final Maps and Results". C.L.Bennett, et al. (Dec 2012) lambda.gsfc.nasa.gov Also found in the Astrophysical Journal Supplement 2013.
2	Although the term 'atomism' is derived from Greek, the first known instances of the theory arise from ancient India and particularly the Ajivika and Carvaka schools of philosophy (c. 600-500BCE). This was later developed in the Nyaya and Vaisheshika schools of Hinduism, and also in Jainism. It is unclear whether the Greeks developed the idea separately or were influenced by Indian culture. The term was coined by the ancient Greek philosopher Democritus (460BCE) and later adopted by the Roman philosopher Lucretius. 'Indian Atomism: History and Sources' M.Gangopadhyaya (1981) Atlantic Highlands, New Jersey: Humanities Press. ISBN 0-391-02177-X. Re:disputed origins – 'Lost Discoveries: The Ancient Roots of Modern Science R.Teresi. (2003) Simon & Schuster. ISBN 0-7432-4379-X.
3	BBC4 "The Secrets of Quantum Physics" Professor J.Al-Khalili University of Surrey. See also "Quantum Life: How physics can revolutionise biology" J.S.Al-Khalili Lecture to the Royal Institution, London 2013. http://richannel.org/jim-al-khalili--quantum-life-how-physics-can-revolutionise-biology Jameel "Jim" Al-Khalili British theoretical physicist, born Bagdhad, Iraq 1962 Currently Professor of Theoretical Physics & Chair of Public Engagement in Science at University of Surrey. Additional Reference for strange Quantum Effects – "Quantum purity: How the big picture banishes weirdness" A. Ananthaswamy New Scientist magazine, page 34-37, 2015 issue 3016.
4	Albert Einstein (1879-1955). A German theoretical physicist who changed the world view of existence with his Special Theory of Relativity (1905) that reconciled Newtonian Mechanics with electromagnetic field theory. He later adapted his theories to accommodate the force of gravity which led to his General Theory of Relativity (1916) and ultimately to the formation of Quantum Theory. His most famous equation is $E=MC^2$ which demonstrates that energy and matter are interchangeable.

5	Although René Descartes (1596 –1650) has been attributed with the saying, "cogito ergo sum", Plato had similar ideas and Aristotle, (in his work Nicomanchean Ethics), described the notion in full. Descartes used the phrase to explore his doubts about religion and the possibility of a deceiving God, before seeing if any of his beliefs could survive his doubts.
6	Technically the section of Idealism which denies physical existence is known as Metaphysical Solipsism – not the most appealing title, possibly by design.
7	"Consciousness Explained" D.Dennett, 1991, p29.
8	Quantum Mechanics is the modern scientific discipline that looks into the operations of existence within atoms, while Newtonian Physics looks at how collections of atoms operate. It arose at the end of the 19th century when certain observations could not be explained by classical Newtonian physics. The term 'Quantum' effectively refers to the minimum amount of any physical entity that is involved in a reaction or 'interaction'. It implies that things can only change in discreet quantities... all be it very small quantities.
8a	Various articles by commentators and top scientists will make the point that Quantum Mechanics remains based in causality even if the nature of the cause is different at the sub-atomic level – for example;- http://crossexamined.org/objection-premise-1-kalam-doesnt-quantum-mechanics-violate-causal-principle/ http://phys.org/news/2015-03-quantum-imply-causation.html http://www98.griffith.edu.au/dspace/bitstream/handle/10072/14521/40586.pdf?sequence=1 https://www.quora.com/Does-quantum-entanglement-violate-causality
9	'Analytic philosophy of religion: its history since 1955' S.Duncan Humanities-Ebooks, also "An Introduction to the Philosophy of Religion" – B.Davies – Oxford University Press.
10	Plato, a student of Socrates popularised a number of concepts about creation and origin - most famously in a work called 'Timaeus' (360BCE) which is a notional dialogue between his mentor Socrates, a character called Timaeus of Locri, plus Hermocrates, and Critias. In this he refers to a 'demiurge' or eternal force for change which would fashion something from eternal base material and therefore that creation must reflect the characteristics of the demiurge (which was not necessarily a being). Ref: ideas on motion Macmillan Encyclopedia of Philosophy (Vol. 2 -1967) - "Cosmological Argument for the Existence of God",

11	The first reasonably accurate calculation of the age of the universe since an event known as the Big Bang, was made by astronomer Alan Sandage in 1958. Originally rejected as being too short a period compared to estimates (at that time) for the oldest stars (then thought to be 25 billion years old), Sandage's estimate of 13.77 billion years was verified by a number of later studies including improved techniques for ageing stars which then showed the oldest observed star to be 13.2 billion years old. "Current Problems in the Extragalactic Distance Scale." A R Sandage, Astrophysical Journal (May, 1958). "Seven-Year Wilson Microwave Anisotropy Probe (WMAP) Observations: Sky Maps, Systematic Errors, and Basic Results" nasa.gov.
12	The term 'Big Bang' was first coined by the English astronomer Sir Fred Hoyle in a radio broadcast in 1949. Hoyle did not support the theory, (favouring an alternate 'steady state' model), and apparently used the term in a jocular and perhaps slightly dismissive fashion.
13	"Un univers homogène de masse constante et de rayon croissant rendant compte de la vitesse radiale des nébuleuses extragalactiques" G. Lemaitre (1927) - Scientific Society of Brussels. "The Evolution of the Universe: Discussion" G. Lemaitre (1931) – Nature
14	"A Relation Between Distance and Radial Velocity Among Extra-Galactic Nebulae." E. Hubble (1929).
15	H.Alpher, R.A.,Bethe, Gamow,G.- The Origin of Chemical Elements (1948) Physical Review
16	The notion of the Big Crunch stems from the work of the Russian physicist /mathematician Alexander Friedmann in 1922 and his a simplified model of the universe based on Einstein's equations (1924 – 'On the possibility of a world with constant negative curvature of space') Scientific theory only has two factors at play in the expansion of the Universe: 1) the force of the explosion and 2) the strength of gravity holding it back - related to the density of matter in space. If the density of material in space makes gravity stronger than the force of the explosion the expansion of the universe will ultimately stop and then reverse.
17	Saul Perlmutter, Brian P. Schmidt, and Adam G. Riess were awarded the Nobel Prize for Physics in 2011 for their discovery that the expansion of the Universe was accelerating. They achieved this by observing that the redshift in type 1a supernovae was increasing.
18	For info about the Planck Space Telescope please see the European Space Agency website http://www.esa.int/Our_Activities/Space_Science/Planck/Science_objectives

19	The 'Faster than Light' experiment performed by Nicolas Gisin (Swiss physicist b.1952) et al (University of Geneva) in 1997 by sending paired particles down long fibre optic cables. When they were over 10 kilometres apart, actions forced onto one particle would be 'instantly' mirrored by the other. A subsequent 2008 experiment calculated the speed of communication to be at least 10,000 times the speed of light. *"Violation of Bell inequalities by photons more than 10 km apart"* W.Tittel, J.Brendel, H.Zbinden, N.Gisin (1998) Physical Review Letters
20	Professor Stephen William Hawking (b.January 8, 1942) is an English theoretical physicist, cosmologist, and author whose work and ideas have been televised on many subjects. He first came to prominence for his theories about Black Holes which are at the centre of every galaxy. In his book 'The Grand Design' he talks about modelling the expansion of the Universe. The book also led to a 2012 TV mini series on Discovery Channel. In it he presents and expands upon theories of time which began in his book "A Brief History of Time"
21	String Theory has been presented in 5 different forms but they all suggest that forces can be viewed as short lines of influence that exist in only one dimension, however when they vibrate they will effectively spread out into two dimensions or more. The theories go on to suggest that such Strings can join together in various ways to create different types of particle and also for a basis for explaining gravity.
22	M-Theory is an extension or amalgamation of the earlier 'String-Theories' and is purported to encompass them all. It was originally proposed by Edward Witten in 1995, the 'M' apparently reflecting a base concept that force could exist as 'membranes' or sheets. Those sheets could be bent to form tubes or spheres that could form the basis of particles. By amalgamating the different String Theories its concept of existence stretches to 10 possible dimensions of existence plus Time, (7 of which we have no current awareness of).
23	"Origins of Existence" F. Adams (2002) Free Press "Dreams of a Final Theory" S.Weinberg. (1993) Hutchinson
24	The German theoretical physicist Werner Karl Heisenberg (5 Dec 1901 – 1 Feb 1976) won the 1932 Nobel Prize in physics for the creation of quantum mechanics. His founding paper was published in 1925 and enhanced with subsequent work in collaboration with Max Born and Pasqual Jordan. He is perhaps most famous for his 'Uncertainty Principle' on which many of his ideas were founded (first published in 1927).
25	The 'Corpuscular Theory of Light' had been proposed by Isaac Newton but the double slit experiment by Thomas Young (1773-1829), first performed in the early 1800s, appeared to disprove it. However the idea of particles was later resurrected due to the discovery of the photoelectric effect in 1887, because individual electrons were observed being generated.

26	"Unsharp particle-wave duality in a photon split-beam experiment". P. Mittelstaedt; A. Prieur, R. Schieder (1987). *Foundations of Physics*
26a	Dual slit experiments were first performed with electron beams in 1961 by Clauss Jonsson at the University of Tubingen. "Zeitschrift für Physik" See also – "Electron diffraction at multiple slits" C.Jönsson (1974). American Journal of Physics. Experiments conducted with single electrons were first performed in 1974 by Italian physicists P.G.Merli, G.F.Missiroli, and G.Pozzi using biprisms instead of slits - "On the statistical aspect of electron interference phenomena". (1976). American Journal of Physics.
26b	"Quantum interference experiments with large molecules" O.Nairz; M.Arndt; A Zeilinger (2003). American Journal of Physics. "Quantum wonders: Corpuscles and buckyballs" 2010 New Scientist: (Subscription needed for full text)
27	"The Feynman Lectures on Physics, Volume III" R.P.Feynman, R.Leighton, M.Sands (1965) Massachusetts: Addison-Wesley
27a	"Complementarity and the Copenhagen Interpretation of Quantum Mechanics" D.Harrison, (2002).Dept. of Physics, U. of Toronto. Retrieved 2008-06-21. "Quantum Mechanics 1925–1927: Triumph of the Copenhagen Interpretation" D.Cassidy (2008). Werner Heisenberg. American Institute of Physics.
27b	"Unsharp particle-wave duality in a photon split-beam experiment" P. Mittelstaedt; A. Prieur; R. Schieder (1987). Foundations of Physics. "Disentangling the wave-particle duality in the double-slit experiment" M.Francis. (2012) Ars Technica.
28	Thomas Warren Campbell, Jr.* (b. Dec. 9, 1944*) physicist and author of the book "My Big Toe" – meaning Theory of Everything, in which he proposes a model of existence similar to a Virtual Reality where physical matter is configured by meta data held outside our 3 dimensions. * - subject to confirmation.
29	An article published by Physics Essays Publication entitled "Consciousness and the double-slit interference pattern : Six experiments" (written by Dean Radin, Leena Michel, Karla Galdamez, Paul Wendland, Robert Rickenbach, and Arnaud Delorme) describes 6 experiments conducted to establish whether the double slit results, (which they refer to as the 'QMP'), could be attributed to the effects of consciousness, or was due to some other factor within equipment or general circumstances, (eg. temperature etc). After hundreds of tests their conclusion was that... "In sum, the results of the present experiments appear to be consistent with a consciousness-related interpretation of the QMP" Article downloaded 28th Jan 2015 from http://www.deanradin.com/papers/Physics%20Essays%20Radin%20final.pdf

30	Bell's Theorem essentially provide a way to test whether traditional views of cause and effect hold true instead of the theory of Quantum Entanglement. It has been described as one of the most profound in science. First published as part of a 1964 paper entitled *"On the Einstein Podolsky Rosen paradox"* (John Stewart Bell b.1928 – d.1990 Northern Irish physicist)
31	Alain Aspect French Physicist b.1947 *"Experimental Tests of Realistic Local Theories via Bell's Theorem"* A.Aspect, P.Grangier; G.Roger (1981) Physical Review Letters *"Experimental Test of Bell's Inequalities Using Time-Varying Analyzers"*, A.Aspect; J.Dalibard; G.Roger (1982) Physical Review Letters
32	www.tudelft.nl/en/current/latest-news/article/detail/einsteins-ongelijk-delfts-experiment-beeindigt-80-jaar-oude-discussie/ Delft University of Technology
33	In the field of quantum mechanics the physicist David Bohm proposed that certain ***illogical*** situations which have emerged from theories and observation could be resolved if 'superpositions' did not always have to occur, and to achieve this he introduced the notion of another variable – a purely physical 'guiding field' or wave - but ***not*** localised in space, and without energy. This adaptation of an earlier 'pilot wave' theory by Louis de Broglie in the 1920s, was proposed in 1952 as part of his 'Causal Interpretation' theory.
34	"Origins of Existence" Fred Adams The Free Press 2002 "A Short History of Nearly Everything" Bill Bryson Transworld 2003
35	*"Potential Matter.—A Holiday Dream"*. A. Schuster Nature **58** (1503): 367. (1898). Bibcode:1898Natur..58..367S. doi:10.1038/058367a0.
36	Supermassive Black Holes are enormous hidden objects, one of which is believed to exist at the heart of every galaxy, (including our own 'Milky Way'), holding it together and causing it to spin. The theory is that huge amounts of material have been gathered together causing gravity to be so strong that even light cannot escape from it. Stars and their planets which are closest to the Black Hole get sucked-in and destroyed, increasing the mass of the object but they also release some energy as beams of radiation from their poles when they're 'feeding', and also separately in the energy required to maintain its force/hold over the spiralling stars. To give you a sense of scale, these objects are thought to exceed the mass of millions, if not billions of our suns, (solar masses).
37	S.W.Hawking, (1974). "Black hole explosions?" *Nature* **248** (5443): 30–31

| 38 | I believe that Paul Steinhardt of Princeton University and Neil Turok of Cambridge University came up with this idea at a conference which they organised in 1999 at Cambridge. The idea describes the curtains as 'branes' (similar to membranes) and suggests that these may exist in other dimensions of existence – per M-Theory.

The full Epkrotic Model of Brane Cosmology was finally proposed by J.Khoury, B. Ovrut, P. Steinhardt and N.Turok in 2001. |
|---|---|
| 39 | As an example, the James Randi Educational Foundation still offers a $1million prize for anyone who can demonstrate paranormal powers under test conditions, however since Randi's death the organization has suspended their application process for the general public as they were apparently getting deluged with bogus claimants, and as at 9/1/2015 were devising new procedures to minimise the administrative burden claiming on their site that "The overwhelming majority refused to fill out the application or even state a claim that can be tested." While we wait for the Foundation to issue a new process any known psychic with an independent TV crew in tow can apparently still get tested. Per Randi.org downloaded 17 Aug 2016. |
| 40 | I believe that this has now been officially proposed by scientists: Neil Turok from Cambridge, Burt Ovrut University of Pennsylvania and Paul Steinhardt from Princeton – partly linked to their Epkyrotic Model – see 38 above |
| 41 | "Dark Energy and the Accelerating Universe". J.A.Frieman; M.S.Turner; D.Huterer. (2008). Annual Review of Astronomy and Astrophysics. 46 (1): 385–432 |
| 42 | Based on an edited transcript of a BBC radio interview with John Bell in 1985 " *The Ghost in the Atom: A Discussion of the Mysteries of Quantum Physics*" by Paul C. W. Davies and Julian R. Brown, 1986/1993, |
| 43 | The "Big Bell Test" experiment involving 100,000 people worldwide was co-ordinated by the ICFO Institute of Photonic Sciences via 12 laboratories across the globe over a 48hour timeframe.

https://www.icfo.eu/newsroom/news/article/3332

http://bist.eu/100000-people-participated-big-bell-test-unique-worldwide-quantum-physics-experiment/ |

Index